The Cardiac Lymphatic System

Ganga Karunamuni
Editor

The Cardiac Lymphatic System

An Overview

Springer

Editor
Ganga Karunamuni
Department of Pediatrics
Case Western Reserve University
Cleveland, OH, USA

ISBN 978-1-4614-6773-1 ISBN 978-1-4614-6774-8 (eBook)
DOI 10.1007/978-1-4614-6774-8
Springer New York Heidelberg Dordrecht London

Library of Congress Control Number: 2013935496

Printed on acid-free paper

Springer is part of Springer Science+Business Media (www.springer.com)

Preface

The heart is invested with a complex, intertwining network of blood and lymphatic vessels which, respectively, provide the cardiac tissue with oxygen and nutrients and eliminate excess fluid from the interstitium. The coronary blood vessels have been the focus of much investigation in the past few decades. On the other hand, the literature regarding the cardiac lymphatic vessels remains sparse, despite their important role in maintaining normal heart function. For example, in the event of lymphatic blockage, destruction, or dysfunction, cardiac edema as well as fibrosis and inflammation can be observed in the affected regions of the heart. With this in mind, a better understanding of the cardiac lymphatic network and its ability to regulate fluid homeostasis within the heart could give us insight into developing therapies for the alleviation of several cardiac pathological conditions.

This book aims to provide in-depth coverage of the cardiac lymphatic vessels and the essential nature of their patterning and development in the heart tissue. The volume is organized into three parts: the anatomy of the cardiac lymphatic system (Chaps. 1 and 2), the cardiac lymphatics and heart disease (Chaps. 3, 4, 5, 6, and 7), and cardiac lymphatic signaling (Chaps. 8, 9, 10, and 11). The first two chapters illustrate the general anatomy of the lymphatic vessels existing within the epicardium, myocardium, and endocardium of the heart. Chapter 3 investigates the role of the lymphatics in the development of congenital heart disease, while Chap. 4 reviews their function in a variety of cardiac pathologies including myocardial infarction and congestive heart failure. Chapter 5 covers the valvular lymphatics under both normal and disease conditions. Chapter 6 discusses some therapeutic applications targeting the cardiac lymphatics during heart disease as well as the limitations of such practices, and Chap. 7 outlines the various imaging modalities that are available for the visualization of the lymphatic vasculature. Chapters 8, 9, and 10 highlight certain crucial signaling pathways (Tie, VEGF, and HIF, respectively) that are involved in the regulation of the cardiac lymphatics. Finally, Chap. 11 delves into the potential of the epicardium to act as a source of stem cells for damaged or regenerating lymphatic vessels.

The volume will be useful to a broad audience interested in cardiovascular medicine and physiology, including clinicians, students, and researchers in the fields of developmental biology, cardiology, and applied anatomy. The groundwork in this book will be able to provide readers with vital information on the crucial role played by the cardiac lymphatic vessels in preserving normal heart function.

My primary thanks go to the contributors, key leaders in the field, without whom this book would not have been possible. Their time and enthusiasm are greatly appreciated. I would also like to thank my family, especially my husband Matthew Krock, for their unwavering love and support during this project.

Cleveland, OH, USA Ganga Karunamuni, PhD

Contents

Contributors

H. Scott Baldwin Division of Cardiology, Vanderbilt Children's Hospital, Nashville, TN, USA

Manish Bansal Department of Pediatrics/Pediatric Cardiology, UH Rainbow Babies and Children's Hospital, Case Western Reserve University, Cleveland, OH, USA

Bharti Shivayogi Bhusnurmath Department of Pathology, St. George's University School of Medicine, St. George, Grenada

Shivayogi Bhusnurmath Department of Pathology, St. George's University School of Medicine, St. George, Grenada

Kathryn Brown MRC Center for Transplantation, King's College London School of Medicine at Guy's, King's and St. Thomas' Hospitals, London, UK

Alexey Dashkevich Transplantation Laboratory, University of Helsinki and Helsinki University Central Hospital, Helsinki, Finland

Galina Dragneva Department of Biotechnology and Molecular Medicine, A.I. Virtanen Institute, University of Eastern Finland, Kuopio, Finland

Lindsey A. Edwards MRC Center for Transplantation, King's College London School of Medicine at Guy's, King's and St. Thomas' Hospitals, London, UK

Jin-Ok Jeong Department of Internal Medicine, Chungnam National University Hospital, Regional Cardiovascular Center, Daejeon, Korea

Ivana Kholová Fimlab Laboratories, Pathology, Tampere University Hospital, Tampere, Finland

Linda Klotz University College of London, Institute of Child Health, Molecular Medicine Unit, London, UK

Karl B. Lemström Transplantation Laboratory and Department of Cardiothoracic Surgery, University of Helsinki and Helsinki University General Hospital, Helsinki, Finland

P. Charles Lin Mouse Cancer Genetics Program, National Cancer Institute, Frederick, MD, USA

Shin Lin Stanford University School of Medicine, Falk Cardiovascular Research Center, Stanford, CA, USA

Marios Loukas Department of Anatomical Sciences, St. George's University School of Medicine, St. George, Grenada

Jörg Männer Department of Anatomy and Embryology, University Medicine Göttingen, Göttingen, Germany

Antti I. Nykänen Transplantation Laboratory and Department of Cardiothoracic Surgery, University of Helsinki and Helsinki University Central Hospital, Helsinki, Finland

Jae-Hyeong Park Department of Internal Medicine, Chungnam National University Hospital, Regional Cardiovascular Center, Daejeon , Korea

Xianghu Qu Department of Pediatric (Cardiology), Vanderbilt University Medical Center, Nashville, TN, USA

Paul Riley Department of Physiology, Anatomy, and Genetics, University of Oxford, Oxford, UK

Stanley G. Rockson Stanford University School of Medicine, Falk Cardiovascular Research Center, Stanford, CA, USA

Sameer Shah Department of Anatomical Sciences, St. George's University School of Medicine, St. George, Grenada

R. Shane Tubbs Department of Pediatric Surgery, Children's Hospital, Birmingham, AL, USA

Jörg Wilting Department of Anatomy and Cell Biology, University Medicine Göttingen, Göttingen, Germany

Wilson Wong MRC Center for Transplantation, King's College London School of Medicine at Guy's, King's and St. Thomas' Hospitals, London, UK

Todd R. Wuest Mouse Cancer Genetics Program, National Cancer Institute, Frederick, MD, USA

Seppo Ylä-Herttuala Department of Biotechnology and Molecular Medicine, A.I. Virtanen Institute, University of Eastern Finland, Kuopio, Finland

Part I
Anatomy of the Cardiac Lymphatic System

Chapter 1
A General Outline of the Cardiac Lymphatic System

Marios Loukas, Sameer Shah, Shivayogi Bhusnurmath, Bharti Shivayogi Bhusnurmath, and R. Shane Tubbs

Abstract The lymphoid system is a collection of freely permeable, thin-walled vessels dispersed throughout the entire body. Lymphatic vessels preserve circulatory homeostasis by clearing blood capillary ultrafiltrate along with excess protein from the interstitium and delivering them back to the venous circulation. Very little is known about the lymphatic system as it relates to the heart, and therefore, the subject is often neglected in textbooks. Many pathological conditions have been linked to obstruction of lymph flow and even, in rare cases, to cardiac lymphoma. The aim of this chapter is to review the anatomy of cardiac lymphatics based on current literature and emphasize the clinical significance of the system.

Keywords Thoracic duct • Lymphatic channels • Lymphatic vessels • Lymph nodes • Lymph drainage • Heart

Introduction

The lymphoid system is a collection of freely permeable, thin-walled vessels dispersed throughout the entire body [1, 2]. The capillary-like lymphatic vessels begin as blind sacs in the interstitial space and have three main functions [1–4]. First, the lymph vessels preserve circulatory homeostasis by clearing blood capillary

M. Loukas (✉) • S. Shah
Department of Anatomical Sciences, St. George's University
School of Medicine, St. George, Grenada
e-mail: mloukas@sgu.edu

S. Bhusnurmath • B.S. Bhusnurmath
Department of Pathology, St. George's University
School of Medicine, St. George, Grenada

R.S. Tubbs
Department of Pediatric Surgery, Children's Hospital, Birmingham, AL, UK

G. Karunamuni (ed.), *The Cardiac Lymphatic System: An Overview*,
DOI 10.1007/978-1-4614-6774-8_1, © Springer Science+Business Media New York 2013

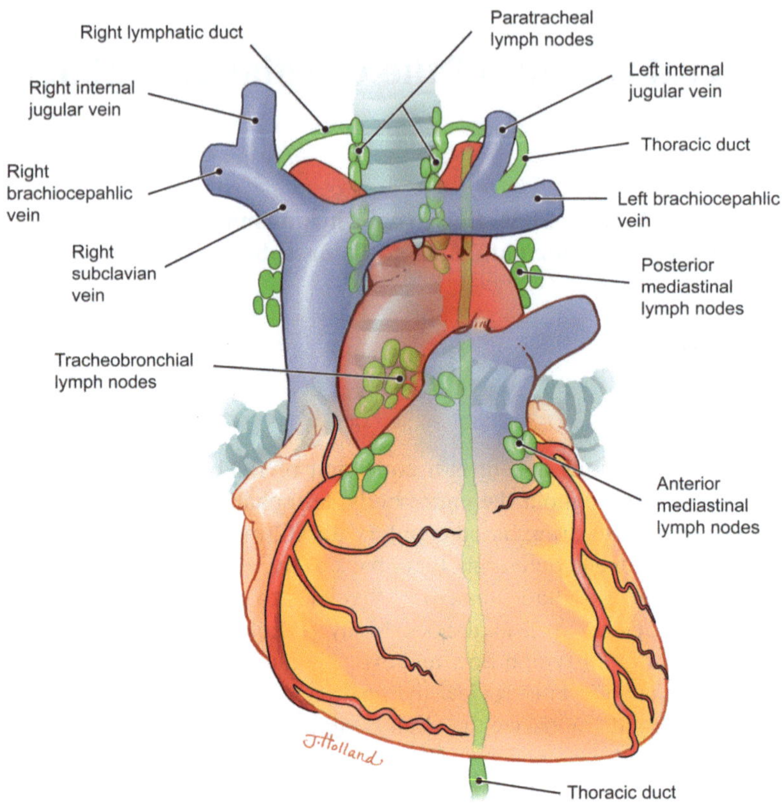

Fig. 1.1 A schematic representation of the lymphatic drainage of the heart

ultrafiltrate along with excess protein within the interstitium and delivering them back to the venous circulation [3, 4]. This clear, protein-rich fluid found in lymphatics is known as lymph [1, 4]. Second, lymphoid tissue in the intestinal tract aids in nutrition by absorbing digested fat [3]. Third, lymph vessels play a primary role in defense. Foreign antigens, bacteria, and lymphocytes found in the interstitium drain with the lymph to regional lymph nodes and trigger activation of the immune system [3]. The diagnosis, prognosis, and treatment of many diseases, not withholding malignancies, require thorough knowledge of lymph circulation and drainage. There are very few studies in the literature on the cardiac lymphatics in particular, and in fact, any mention of cardiac lymph drainage in textbooks is rare. This study focuses on lymphatics as they relate to the heart.

Lymphatics in the heart can be divided into three categories based on their size: minute valveless capillaries, medium lymphatic collecting vessels (usually valveless), and large valved lymphatic trunks [5]. In general, lymph flows from the subendocardium outward toward the subepicardial lymphatic plexus [1, 6, 7]. These plexuses feed into larger lymphatic vessels that follow the path of the main coronary blood vessels in the anterior and posterior interventricular grooves and the coronary sulcus (Fig. 1.1) [1, 5, 7]. Kline [5] observed that these lymphatic vessels within the epicardium resembled white bands. Many pathological conditions have been linked to obstruction of lymph flow and even, in rare cases, to cardiac lymphoma [8–10]. The aim of this chapter is to review the current literature associated with the cardiac lymphatic system, with an emphasis on its clinical significance.

History

Many authors credit Rudbeck as the first researcher to bring attention to the lymphatics in the heart [11, 12]. In 1653, Rudbeck [13] described the subepicardial lymph vessels of a dog, as reported by Patek [12]. The heart of a dog is often used as a surrogate model for the human heart when studying lymphatics because the two are nearly identical in anatomy [14]. Since Rudbeck, several anatomists have studied these lymphatics in more detail. According to Bradham et al. [15], Nuck discovered several decades later that mercury could be introduced directly into the cardiac lymph vessels. In 1924, Aagaard [16] discovered lymphatic plexuses in the subendocardial, myocardial, and subepicardial tissue layers after injecting dye into the heart muscle. In 1939, Patek [12] substantiated this finding using India ink. Patek [12] observed that lymph is drained outward in valveless channels from the subendocardial plexus through to the myocardial plexus and subepicardial plexuses and is finally collected into a large single trunk draining the entirety of the heart [12]. The following year, Drinker et al. [17] cannulated a major efferent lymphatic trunk to measure the flow, pressure, and composition of lymph leaving the heart.

Since 1940, several studies have confirmed these preceding basic anatomical findings. In 1966, Johnson and Blake [18] utilized hydrogen peroxide to distend the lumens of the cardiac lymphatic vessels and blood capillaries in an effort to investigate morphological differences, and in the process they demonstrated a vast network of subendocardial and subepicardial lymphatics [18]. In a study involving mongrel dogs, Bradham et al. [15] confirmed that a vast network of lymphatics is ubiquitous in the ventricular myocardium, even more so on the left side (Fig. 1.2). In 1982, Miller [19] outlined the path of cardiac lymph drainage from the major collecting trunks to the mediastinal lymph nodes, with the lymph finally emptying into the thoracic duct.

Embryological Studies

With his study of embryos and fetuses of humans, Kampmeier [20] utilized vital dye to show that lymph vessels of the heart originated from two plexuses. The first plexus began in the right ventricle, followed the right coronary artery proximally, ascended between the aorta and pulmonary trunk, and finally drained into the

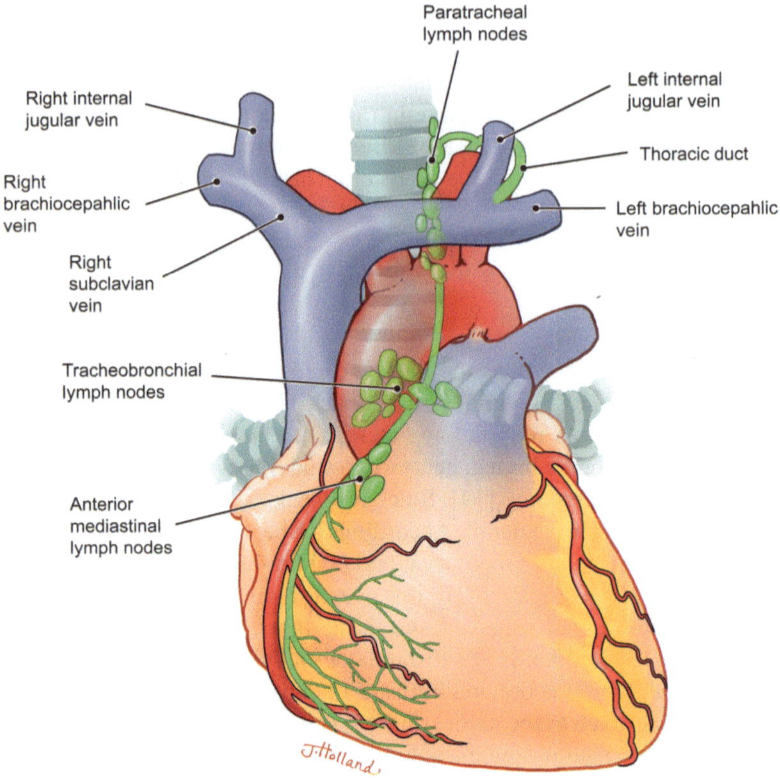

Fig. 1.2 A schematic representation of the right lymphatic drainage of the heart

upper thoracic duct adjacent to the left jugular lymph sac. The second plexus was observed to originate in the left ventricle, course proximally along with the left coronary artery, and drained into the right jugular sac (deriving from the pretracheal plexus) [20]. Kampmeier [20] inferred that these discrete efferent lymphatic paths persisted in adults. However, Feola et al. [21] disagreed with this conclusion. In their study, seven out of nine cadaveric specimens exhibited the left and right major lymphatic trunks unifying as one main lymphatic vessel superior to the heart [21]. In one of the earliest studies on lymphatic origin, Cash [22] demonstrated in embryo pigs that there is initially one large lymphatic plexus covering the entire surface of the heart, with vessels penetrating and networking deeper during late development.

In animal studies, the atrioventricular (AV) valves of both adult dogs and puppies have been observed to have lymphatic capillaries [23]. The capillaries visualized in the adults were more developed, which supported the theory that lymphatic vessels of the heart continue to develop into adulthood [23]. In contrast, human cardiac lymphatics have been demonstrated to conclude development in early stages of fetal growth [24].

The Subendocardial Plexus

As stated above, the network of lymphatic vessels in the subendocardium are extensive [18]. Uhley et al. [25] observed the branches of these vessels crossing at myriad angles. It has been reported that small lymphatic capillary systems, such as the subendocardial plexus, originate from the blood endothelium and basement membrane and are observed as sizeable dilatations [26]. Lupinski [26] approximated that 1,300 blood capillaries drain into each lymphatic capillary. As discussed, lymph flow in the heart progresses outward from the subendocardial plexus to the myocardial plexus [12]. Lupinski [26] opined that there is no clinical significance to this pattern of outward flow.

Myocardial Plexus

Studies detailing the myocardial plexus are scarce. Anatomy and pathology textbooks forego the topic all together [27], or else only note that they exist [1]. Nevertheless, Bullon and Huth [28] described the intricate anatomy of lymphatic vessels in the myocardium. They found that myocardial lymphatic capillaries were sparse compared to blood capillaries, and it was noted that a continuous basal membrane was absent [28]. Bullon and Huth [28] suggested these lymphatics possessed the ability to adjust immediately to an increased functional load as a result of the multitude of cellular junctions encompassing the vessels. In addition to blood capillaries, Bullon and Huth [28] reported that the basement membranes of nerves, fibrocytes, and myocytes were in direct contact with the adjacent myocardial lymphatic vessels.

Recent research by Lupinski [26] has expanded on prior knowledge of the myocardial lymphatic plexus. He described the numerous lymphatic capillaries as "basket-like" and confirmed they are positioned adjacent to blood capillaries in the interstitium [26]. Lymph from the myocardial plexus collects into channels, which drains to the subepicardial plexus via channels in the connective tissue septum [26].

Subepicardial Plexus

Patek [12] discovered the subepicardial plexus network of lymphatic capillaries, with accompanying collecting channels between the myocardial and epicardial muscle layers, by injecting India ink into the subepicardium. According to Patek [12], lymphatics can be classified into five types that are determined by the region being drained. The first type includes all of the lymphatics initially flowing into the capillary plexuses, and the second type involves the lymphatic vessels that make up the capillary plexus. As the diameter of the lymphatic collecting vessels increase, the type numbers increase correspondingly with type five being the largest lymphatic trunk [12]. Patek's [12] results were substantiated by Johnson and Blake [18], in humans, as well as in other species. In addition, Johnson and Blake [18] observed the efferent lymphatic vessels flowing into larger lymphatic trunks in the AV grooves and finally emptying within the main lymphatic collecting trunk of the heart. More recent studies also support their findings [26].

Pericardial Lymphatics

Prior to 1977, the consensus was that the right lymphatic duct drained the majority of cardiac lymph. However, Leeds et al. [29] observed multiple pathways of lymphatic drainage from the parietal pericardium into both the right and thoracic lymphatic ducts of canines using radiolabeled lymph. Following the canine pericardial lymph drain more proximally, Miller et al. [30] showed the lymphatic ducts emptying into the left and right upper mediastinal nodes and further to the parasternal internal thoracic chain of lymphatics bilaterally.

In humans, lymph in the pericardial space has been shown to drain by several different pathways [31]. Anteriorly, lymph from the upper aspect drained toward the phrenic neurovascular bundle, turned superiorly to the brachiocephalic veins, and ultimately emptied into both anterior mediastinal lymph nodes (Fig. 1.1). Lymph from lower regions either drained toward the phrenic neurovascular bundle and then moved inferiorly to enter the diaphragm or coursed ventrally, entering the prepericardial lymph nodes at the juncture between the diaphragm and pericardium. Laterally, lymph from the upper aspect of the pericardium mostly flowed superiorly into the peribronchial and tracheobronchial lymph nodes. The lower part mirrored that of the anterior pericardial space. A posterior part of the lateral pericardium drained to a chain of lymph nodes ventral to the esophagus and dorsal to the inferior vena cava. Lymphatics from the posterior pericardium, including the cupula, ultimately drained superiorly to the tracheobronchial lymph nodes. Finally, lymph from the diaphragmatic pericardium followed several different pathways. The right side drained into the right pericardial lymph node, the ventral aspect into the prepericardial nodes, and the central aspect into the posterior mediastinal or tracheobronchial lymph nodes (Figs. 1.2 and 1.3) [31]. In addition, Eliskova et al. [31] observed a second layer of lymphatics within the adipose and areolar tissues.

Lymphatics of the Cardiac Valves

The presence of lymphatics within AV valves is a point of contention [32]. However, Miller et al. [33] demonstrated lymphatic vessels in the valves of live canines. In 1988, Noguchi et al. [23] utilized various methods, including light and electron microscopy, to substantiate this finding. Within the atrial region of the leaflets of all

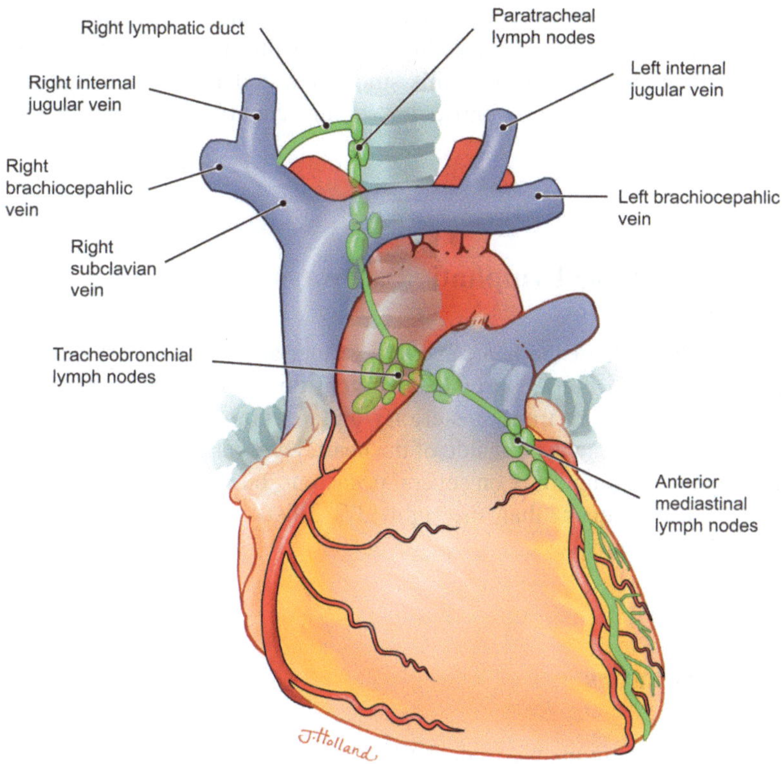

Fig. 1.3 A schematic representation of the left lymphatic drainage of the heart

the valves, especially the mitral valve, they discovered subendocardial lymph-filled capillaries. Noguchi et al. [23] found the majority of the lymphatic vessels of the leaflets were more proximal, diminishing as they approached the free border. The following year, Ichikawa et al. [34] observed the lymphatic vessels originating as a single endothelial cell projecting into the chordae tendineae from the papillary muscles. Miller et al. [33] suggested that it is highly probable that the pathways of lymphatics of the AV valves are connected with the corresponding aortic and pulmonary valves.

Right-Sided Cardiac Lymphatic Pathways

Riquet and Hidden [35] investigated lymphatics of the right atrium and ventricle in cadaveric specimens. After dye was injected at the right cardiac border, they observed lymph flow to the main right efferent lymphatic collecting vessel and continue superiorly toward the left anterior mediastinal lymphatic chain.

Riquet and Hidden [35] introduced dye within the interventricular groove and observed it drain into the thoracic duct, passing through the right paratracheal lymph nodes on the way. Prior to this study, it was demonstrated that the right subendocardial aspect of the ventricular septum contained the more substantial lymphatic network [36].

Left-Sided Cardiac Lymphatic Pathways

Studying the left side of the heart as well, Riquet and Hidden [37] again utilized dye, this time within the left principal efferent ducts. They demonstrated lymph traversing the pulmonary trunk posteriorly and draining into the right pretracheal lymphatic chain, eluding the left side of the heart. Dye was also reported to enter the right principal collecting trunk on several occasions, but generally, variations of the left side are less frequent than the right [37]. Eliskova et al. [38] found similar results in Macaque monkeys.

Atrial Lymphatics

Previously, many were unable to detect lymph capillaries in the atria. More recent studies utilizing light and electron microscopy have shown scant lymphatics in the subepicardium of atria [39, 40]. Marchetti et al. [40] opined that lymphatic vessels have only been noted in the subepicardium of the atrium as a consequence of its very thin walls.

Ventricular Lymphatics

As one would expect, lymphatic vessels in the ventricles are consistent with the organization of subendocardial, myocardial, and subepicardial plexuses, as noted above [36, 41]. In a study in canines, Uhley and Leeds [42] showed that papillary muscles contained an extensive lymphatic system. Burch et al. [43] suggested that disruption of the lymphatics in papillary muscles would cause derangement of the mitral valvular complex and therefore malfunction of the mitral valve. Uhley and Leeds [42] observed an intricate reticulation of lymph capillaries articulating the surface of both papillary muscles of the mitral valve. In addition, they demonstrated thin spikes of lymphatic vessels projecting down the chordae tendineae [42].

Lymphatics Related to the Cardiac Conduction System

Studies have shown a significant anatomical association of the lymphatic capillaries in the subendocardium with the right bundle branch [25]. Uhley et al. [25] suggested that pathology related to cardiac lymph flow may delay or even block conduction along the fascicle, with an endpoint of fibrosis. Uhley et al. [25] speculated that the AV node region drained lymph from the inferior part of the chambers on the right side of the heart. They proposed that damaged tissue released large amounts of potassium that could reach the AV node in this fashion and disrupt its extracellular environment, which could explain the transient manifestations of AV conduction delays associated with an inferior infarction [25]. Uhley et al. [25] also described an anterior septum infarction leading to potassium-rich lymph draining to the right bundle, delaying conduction and creating the possibility of a complete bundle branch block.

Golab [44] demonstrated that lymph originating in the conduction system drains via the myocardial plexus. He found that most of the lymphatics from the AV node emptied directly into the main left lymphatic trunk and the majority of lymph collected from the sinoatrial node flowed directly into the main right lymphatic trunk [44].

Etiology and Sequelae of Disruption to Cardiac Lymph Flow

As stated above, lymphatics play a key role in clearing protein-rich fluid from the interstitium [3, 4]. Inhibition of lymph flow elsewhere in the body can lead to serious complications, and it is no different in the heart. Disruption of cardiac lymphatics was shown to cause subsequent build up of excess interstitial fluid and proteins [45]. This lymphedema has been found to initiate many pathological processes, such as interstitial fibrosis and pericardial effusion, thus impairing cardiac function [46, 47]. It is important to consider the etiology and sequelae of the obstruction of lymph flow in the heart because they are clinically significant.

The literature supports many different sources of impairment to lymph flow, which could be roughly separated into pathological and iatrogenic origins. We propose the pathological causes could be further divided into three groups: those causing lymph stasis due to decreased myocardial contractility, those causing lymphatic obstruction by compression of the vessels, and those damaging lymphatics through inflammation and fibrosis.

The flow of cardiac lymph is completely dependent on the contractions of the heart [3, 7, 48]. Myocardial ischemia, increased superior vena caval pressure, ventricular fibrillation, and pulmonary artery hypertension have all been shown to decrease cardiac contractility and are associated with compromised cardiac lymph flow [5, 19, 49]. In contrast, Feola et al. [50, 51] noted an increase in lymph flow following acute myocardial ischemia in canines. They suggested cardiac lymphatics were essential in the reduction of ischemic damage by decreasing interstitial edema and anaerobic tissue metabolites [50, 51]. Cui [52, 53] also supported the benefits of improved cardiac lymph flow ensuing myocardial infarction.

Tumors of the heart have also been shown to disrupt lymph flow by mechanical compression of the vessels. It was reported that metastatic cardiac tumors were a more common cause of mass effect on lymphatics [49]. Primary cardiac tumors, as discussed later, are rare.

Another source of injury to lymphatics is inflammation. Lupinski [54] reported on mediastinal lymph nodes that were damaged after infection, causing blockage of distal lymph flow, and injury to the cardiac lymphatic vessels. The author suggested rheumatic arteritis, endocarditis, syphilis, and tuberculosis as sources of inflammation. On the other hand, the reverse has also been shown to be true. Maurice et al. [55] argued that impaired lymph drainage renders the heart more susceptible to inflammation and infection. Miller [14] provided support to this study and added that cardiac lymphedema predisposed the tissue to fibrosis and fibroelastosis. Several studies, human and animal, have substantiated the idea of lymphatic obstruction initiating a cascade of significant fibroblastic proliferation leading to endocardial and interstitial fibrosis [5, 56–59].

Many have commonly accepted that the mediastinal lymphatics are nonessential with no harmful consequences to their abolition, but the proposed evidence has opposed that theory. Several cardiac surgical procedures, such as valve replacement and cardiac transplant, result in removing or scarring of mediastinal and intrathoracic lymphatic vessels [46]. Lupinski [26] described the destruction of cardiac lymphatic vessels during open-heart surgery as a result of cross-clamping. The ramifications of cardiac lymphatic obstruction have been thoroughly outlined above. Kong et al. [59] suggested lymphatic obstruction is rarely avoidable in cardiac transplantation and that the root of allograft failure was a result of injury to the myocardium from the obstruction. Mehlhorn [48] reported that dysfunction of the left ventricle and myocardial edema after surgery contributed to the compromised lymphatic flow.

Lymphoma of the Heart

Malignant cardiac lymphoma is an infrequent, but significant cause of cardiac lymphatic obstruction. The heart is not often initially involved at the time of diagnosis, but metastases of lymphomas to the heart have been observed in 20 % of cases, based on autopsy reports [8, 9, 60]. The prevalence of primary cardiac lymphoma is extremely low, constituting only 1.3 % of all heart tumors and 0.5 % of all extranodal lymphomas [61]. Primary cardiac lymphoma is commonly defined as a non-Hodgkin lymphoma presenting exclusively in the heart, or having the majority of the tumor mass in the heart or pericardium [8, 60]. Jeudy et al. [61] suggested that an increasing prevalence of cardiac lymphomas in transplant recipients and acquired immunodeficiency syndrome patients is a result of the Epstein-Barr virus.

Cardiac lymphomas are most commonly found in, but are not limited to, the right-sided chambers of the heart [8, 60], and the most common sign observed with the lymphomas are pericardial effusions [9]. Because patients with severe disease can be asymptomatic, and clinical suspicion is usually low with imaging findings being nonspecific, cardiac lymphomas tend to be more advanced at the time of diagnosis and therefore are generally associated with a poor prognosis [9, 60, 62]. Widespread dissemination and underlying immunodeficiency also add to the grave outcome [9, 60]. Polychemotherapy including anthracycline was shown to increase survival, with combination radiation and chemotherapy being the most effective [60].

Conclusion

In spite of the fact that most textbooks neglect the subject, the cardiac lymphatic system may play a significant role in the pathophysiology of many diseases. Furthermore, a clear understanding of the anatomy and physiology of cardiac lymphatics may be an important part of successful thoracic operations including improved postoperative outcomes. Although further studies are required, the authors hope this distillation of current knowledge serves as an overview of the cardiac lymphatic system for both the morphologist and the clinician.

References

1. Moore KL, Dalley AF, Agur AMR (2010) Clinically oriented anatomy, 6th edn. Lippincott Williams and Wilkins, Baltimore, MD, p 43
2. Berne RM, Levy MN (1992) Cardiovascular physiology: the microcirculation and lymphatics. Mosby Year, St. Louis, MO, p 168
3. Levick JR (1990) An introduction to cardiovascular physiology: lymph and the lymphatic system. Butterworth Heinemann Ltd., Oxford, pp 159–163
4. Smith JJ, Kampine JP (1990) Circulatory physiology: microcirculation and the lymphatic system, 3rd edn. Williams and Wilkins, Baltimore, MD, pp 136–137
5. Kline IK (1969) Lymphatic pathways in the heart. Arch Pathol 88:638
6. Hollinshead WH (1956) Anatomy for surgeons, vol 2. Paul B. Hoeber, Inc., New York
7. Alexander RW, Schlant RC, Fuster V (eds) (1998) Hurst's the heart, arteries and veins: anatomy of the heart, vol 1, 9th edn. McGraw Hill, New York, pp 52–53
8. Ceresoli GL, Ferreri AJ, Bucci E, Ripa C, Ponzoni M, Villa E (1997) Primary cardiac lymphoma in immunocompetent patients: diagnostic and therapeutic management. Cancer 80:1497–1506
9. Gill PS, Chandraratna PA, Meyer PR, Levine AM (1987) Malignant lymphoma: cardiac involvement at initial presentation. J Clin Oncol 5:216–224
10. Zaharia L, Gill PS (1991) Primary cardiac lymphoma. Am J Clin Oncol 14:142–145
11. Miller AJ (1963) The lymphatics of the heart. Arch Intern Med 112:501–511
12. Patek PR (1939) The morphology of the lymphatics of the mammalian heart. Am J Anat 64:204–239
13. Rudbeck O (1953) [Nova exervitatio anatomia exhibens ductus hepatico aquosos et vasa glandularum serosa, nune primum inventa, aeneisque figuris delineate]. Lauringerus, Euchar. Latin.

14. Miller AJ (2011) The grossly invisible and generally ignored lymphatics of the mammalian heart. Med Hypotheses 76:604–606
15. Bradham RR, Parker EF, Barrington BA Jr, Webb CM, Stallworth JM (1970) The cardiac lymphatics. Ann Surg 171:899–902
16. Aagaard OC (1924) Les Vaisseaux Lymphatique du Coeur Chez l'Homme et Chez Quelques Mammiferes. Levin, and Munkegaard, Copenhagen
17. Drinker CK, Warren MF, Maurer FW, McCarrel JD (1940) The flow, pressure, and composition of cardiac lymph. Am J Physiol 130:43
18. Johnson RA, Blake TM (1966) Lymphatics of the heart. Circulation 33:137–142
19. Miller AJ (1982) Lymphatics of the heart. Raven, New York
20. Kampmeier OF (1928) On the lymph flow of the human heart, with reference to the development of the channels and the first appearance, distribution, and physiology of their valves. Am Heart J 4:210
21. Feola M, Merklin R, Cho S, Brockman S (1977) The terminal pathway of the lymphatic system of the human heart. Ann Thorac Surg 24:531–536
22. Cash JR (1917) On the development of the lymphatics in the heart of the embryo pig. Anat. Rec., 13: 451–464. doi:110.1002/ar.1090130706
23. Noguchi T, Shimada T, Nakamura M, Uchida Y, Shirabe J (1988) The distribution and structure of the lymphatic system in dog atrioventricular valves. Arch Histol Cytol 51:361–370
24. Riquet M, Briere J, Dupont P, Pennhouat G, Hidden G (1993) The embryonic and early fetal development of the lymphatics of the heart and lungs in humans. Surg Radiol Anat 15:369–370
25. Uhley HN, Leeds SE, Sung MA, Patel S, Lobos E (1972) The subendocardial lymphatics of the canine heart. Am J Cardiol 29:367–371
26. Lupinski RW (2009) Aortic fat pad and atrial fibrillation: cardiac lymphatics revisited. ANZ J Surg 79:70–74
27. Gould SE (1960) Pathology of the heart, 2nd edn. Charles C. Thomas, Springfield, IL
28. Bullon A, Huth F (1972) Fine structures of lymphatics in the myocardium. Lymphology 5:42–48
29. Leeds SE, Uhley HN, Meister RB, McCormack KR (1977) Lymphatic pathways and rate of absorption of ^{131}I-albumin from pericardium of dogs. Lymphology 10:166–172
30. Miller AJ, DeBoer A, Pick R, Van Pelt L, Palmer AS, Huber MO (1988) The lymphatic drainage of the pericardial space in the dog. Lymphology 21:227–233
31. Eliskova M, Eliska O, Miller AJ (1995) The lymphatic drainage of the parietal pericardium in man. Lymphology 28:208–217
32. Johnson RA (1969) The lymphatic system of the heart. Lymphology 2:95–108
33. Miller AJ, Pick R, Katz LN (1961) Lymphatics of the mitral valve of the dog. Circ Res 9:1005–1009
34. Ichikawa S, Uchino S, Hirata Y (1989) Lymphatics of the cardiac chordae tendineae with particular consideration of their origin. Lymphology 22:123–131
35. Riquet M, Hidden D (1991) Lymphatic drainage of the right atrium and ventricle of the heart. Surg Radiol Anat 13:235–237
36. Marchetti C, Poggi P, Calligaro A, Casasco A (1985) Lymph vessels of the rabbit heart: distribution and fine structure in ventricles. Lymphology 18:90–95
37. Riquet M, Hidden D (1991) Lymphatic drainage of the left atrium and ventricle of the heart. Surg Radiol Anat 13:238–240
38. Eliskova M, Eliska O, Miller AJ, Palmer A, DeBoer A, Usman Z (1992) The efferent cardiac lymphatic pathways in the Macaque monkey. Lymphology 25:69–74
39. Shimada T, Morita T, Oya M, Kitamura H (1990) Morphological studies of the cardiac lymphatic system. Arch Histol Cytol 53:115–126
40. Marchetti C, Poggi P, Calligaro A, Casasco A (1986) Lymph vessels of the rabbit heart: distribution and fine structure in atria. Lymphology 19:33–37

41. Takada K, Otsuki Y, Magari S (1991) Lymphatics and pre-lymphatics of the rabbit pericardium and epicardium with special emphasis on particulate absorption and milky spot-like structures. Lymphology 24:116–124
42. Uhley HN, Leeds SE (1976) Lymphatics of the canine papillary muscles. Lymphology 9:72–74
43. Burch GE, Depasquale NP, Phillips JH (1963) Clinical manifestations of papillary muscle dysfunction. Arch Intern Med 112:112–117
44. Golab B (1977) Lymphatic vessels of the conducting system of the human heart. Folia Morphol 4:317–322
45. Caro DM, Berjon A, Teijerira J, Duran CG (1972) [Anatomical study of the cardiac lymphatics. Their pathogenic role in pericardial effusions]. Ann Chir Thorac Cardiovasc 11:373–376, French
46. Cui Y, Urschel JD, Petrelli NJ (2001) The effect of cardiopulmonary lymphatic obstruction on heart and lung function. Thorac Cardiovasc Surg 49:35–40
47. Ishikawa Y, Akishima-Fukasawa Y, Ito K, Akasaka Y, Tanaka M, Shimokawa R et al (2007) Lymphangiogenesis in myocardial remodeling after infarction. Histopathology 51:345–353
48. Mehlhorn U, Davis KL, Burke EJ, Adams D, Laine GA, Allen SJ (1995) Impact of cardiopulmonary bypass and cardioplegic arrest on myocardial lymphatic function. Am J Physiol 268:H178–H183
49. Huth F (1983) Disease of cardiac lymph vessels. In: Silver MD (ed) Cardiovascular pathology, vol 1. Churchill Livingstone, New York, Edinburgh, pp 171–190
50. Feola M, Glick G, Pick R (1973) Interrelations between cardiac lymph and experimental infarction in dogs. Clin Res 21:418
51. Feola M, Glick G (1975) Cardiac lymph flow and composition in acute myocardial ischemia in dogs. Am J Physiol 229:44–48
52. Cui Y (2010) The role of lymphatic vessels in the heart. Pathophysiology 17:307–314 [Epub ahead of print]
53. Cui Y (2010) Impact of lymphatic vessels on the heart. Thorac Cardiovasc Surg 58:1–7
54. Lupinski RW (2004) Lymphatic aneurysm of the heart. Lymphology 37:141–150
55. Maurice SM, Palmer AS, Miller AJ, Greene R (2001) Lymphatic drainage of the heart in the laboratory rat. Lymphology 34:145–148
56. Miller AJ, Pick R, Katz LN (1963) Ventricular endomyocardial pathology following chronic impairment of cardiac lymph flow in the dog. Br Heart J 25:182–190
57. Kline IK, Miller AJ, Pick R, Katz LN (1964) The relationship between human endocardial fibroelastosis and obstruction of the cardiac lymphatics. Circulation 30:728–735
58. Gibbs NM, Haworth JC, Rendle-Short J (1957) Endocardial fibroelastosis with generalized lymphadenopathy, oedema, and rash. Br Heart J 19:193
59. Kong XQ, Wang LX, Kong DG (2007) Cardiac lymphatic interruption is a major cause for allograft failure after cardiac transplantation. Lymphat Res Biol 5:45–47
60. Gowda RM, Khan IA (2003) Clinical perspectives of primary cardiac lymphoma. Angiology 54:599–604
61. Jeudy J, Kirsch J, Tavora F, Burke AP, Franks TJ, Mohammed TL et al (2012) From the radiologic pathology: archives cardiac lymphoma: radiologic-pathologic correlation. Radiographics 32:1369–1380
62. Tanaka T, Sato T, Akifuji Y, Sakamoto M, Shio H, Ueki J et al (1996) Aggressive non-Hodgkin's lymphoma with massive involvement of the right ventricle. Intern Med 35:826–830

Chapter 2
Development and Patterning of the Cardiac Lymphatic Network

Jörg Wilting and Jörg Männer

Abstract An overview of the prenatal development of the cardiac lymphatic system can be achieved only when it is seen in the general context of prenatal morphogenesis of the heart. Moreover, since it was found that the development of the cardiac lymphatic network follows, in time and place, that of the coronary blood vessels, cardiac lymphangiogenesis cannot be fully understood without knowledge about the essential steps in the development of the blood supply of the developing heart. We therefore start this chapter with short descriptions of both the morphogenesis of the human embryonic heart and the development of its blood supply, and we will then focus on the development of the cardiac lymphatic system.

Keywords Heart development • Cardiac looping • Trabecular myocardium • Proepicardium • Vasculogenesis • Subendocardial lymphatic plexus • Myocardial lymphatic plexus • Subepicardial lymphatic plexus • Jugular lymph sac • Lymphangiogenesis

Introduction

The heart is the first organ to function in vertebrate embryos. The human heart, for example, starts its pumping action around the twenty-first embryonic day. From its first until its last beat, cardiac pumping function depends on sufficient blood supply

J. Wilting (✉)
Department of Anatomy and Cell Biology, University Medicine Gottingen,
Kreuzbergring 36, Goettingen 37075, Germany
e-mail: joerg.wilting@med.uni-goettingen.de

J. Männer
Department of Anatomy and Embryology, University Medicine Gottingen,
Kreuzbergring 36, Goettingen 37075, Germany
e-mail: jmaenne@gwdg.de

G. Karunamuni (ed.), *The Cardiac Lymphatic System: An Overview*,
DOI 10.1007/978-1-4614-6774-8_2, © Springer Science+Business Media New York 2013

to its myocardium. In the primitive embryonic heart and in the mature heart of several lower vertebrate species, myocardial blood supply comes exclusively from the endocardial lumen. The ventricular myocardium of these hearts has a spongy architecture with intertrabecular spaces, which become filled and emptied with blood during ventricular diastole and systole, respectively. The formation of a thick layer of compact ventricular myocardium, which is a characteristic feature of higher vertebrate hearts, is accompanied by the establishment of a new mode of myocardial blood supply via a system of blood vessels called the coronary vasculature. Due to the fatal consequences of coronary vessel insufficiency (e.g., myocardial infarction), this vascular system receives enormous interest, whereas the second vascular system, the cardiac lymphatics, is mostly neglected, although cardiac lymphedema has also been recognized as a clinically important problem for a long time. In the mature human heart, plexuses of initial lymphatics are found in subendocardial and subepicardial positions, connected by myocardial lymphatics, usually accompanying the coronary arterioles. The lymphatic collectors contain valves and, arranged in pairs, follow the subepicardial branches of the coronary arteries. They drain via the left coronary trunk into the right venous angle and via the right coronary trunk into the left venous angle. The coronary blood vessels are derived from a primarily extracardiac population of cardiac progenitor cells, the proepicardium (PE), which is located near the venous pole of the tubular embryonic heart. PE cells colonize the initially naked myocardial surface of the embryonic heart and give rise, first, to the epicardial mesothelium and, later, to various other cell types, including fibroblasts, endothelial cells, and vascular smooth muscle cells. Embryonic development of the cardiac lymphatics starts with the formation of subepicardial vessels, which proceeds from the base to the apex of the heart. Despite their intimate topographical relationship to coronary arteries, fate-mapping studies have shown that the cardiac lymphatic endothelium does not derive from the PE. The precise origin of the cardiac lymphatics is a current matter of research. It is likely that they derive from sprouts of the jugular lymph sacs (JLS), which are located at the confluence of the cranial and caudal cardinal veins into the common cardinal vein. The co-option of lymphangioblasts into growing cardiac lymphatics has not been studied experimentally. The embryonic origin of cardiac lymphatics and their functions in the pre- and postnatal heart is an open field for future investigations.

Short Overview on the Morphogenesis of the Human Embryonic Heart

The development of the human heart starts during the third week of prenatal development (fifth postmenstrual week). A bilaterally paired *heart-forming field* forms within the splanchnic layer of the lateral plate mesoderms of the tri-laminar embryonic disc. The paired heart-forming field harbors the progenitor cell populations for the myocardial and endocardial cell lineages of the future heart. Experimental and molecular data from nonhuman embryos have shown that each half of the paired heart-forming field can be further subdivided at least into two genetically different

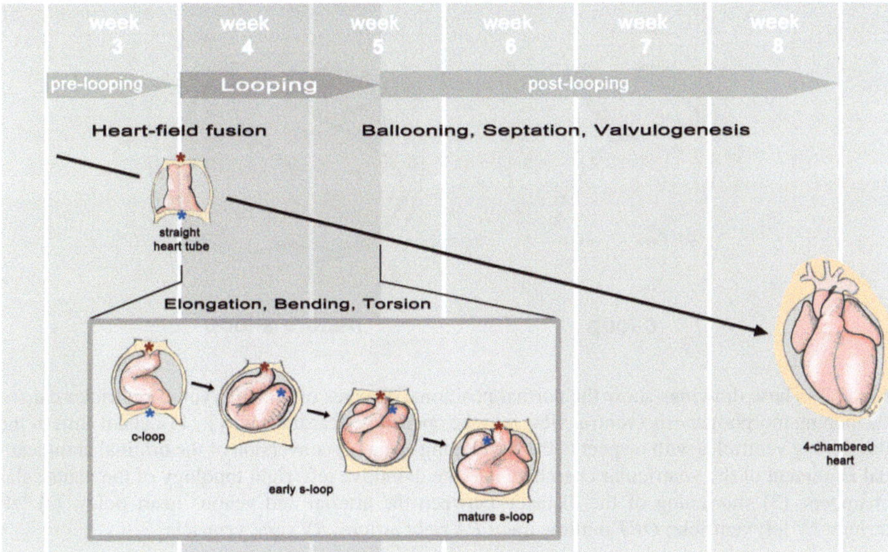

Fig. 2.1 This schematic drawing gives an overview on the morphogenesis of the human embryonic heart. Cardiac morphogenesis can be subdivided into three subsequent phases: (1) *Pre-looping morphogenesis* (third week of prenatal development). A bilaterally paired heart-forming field forms within the lateral plate mesoderm. Heart fields start fusion in the body midline and form a straight heart tube, which is connected to the veins at its caudal end (*blue asterisk*) and to the arteries at its cranial end (*red asterisks*). (2) *Looping morphogenesis* (fourth and fifth week). The tubular heart elongates by addition of new material from the heart-forming field to its cranial and caudal ends. Due to the limited space within the pericardial cavity, the elongating heart tube undergoes bending and torsion and thereby acquires the configuration of a helically wound heart loop. Looping morphogenesis brings the segments of the heart tube and the developing great vessels into an approximation of their definitive topographical relationships. (3) *Post-looping morphogenesis* (fifth to eighth week). The helically wound heart tube becomes transformed into a four-chambered heart by mainly three processes (a) ballooning, (b) septation, and (c) valvulogenesis. By the end of the embryonic period (end of the eighth week), the heart has acquired the basic four-chambered design of the mature human heart

subdomains which have been named the first and second heart fields [1]. For reasons of simplicity, however, in our short description of cardiac morphogenesis, we will not distinguish between various subdomains of the heart field and, therefore, use the term *heart-forming field* in a very general sense.

The right and left halves of the heart-forming field begin to fuse in the embryonic body midline by the end of the third week of development. Thereby, a straight heart tube is formed in front of the developing foregut (Fig. 2.1). It has a single endocardial lumen, which is connected to the veins at its caudal end (venous heart pole) and to the arteries at its cranial end (arterial heart pole). It is frequently stated in the literature that the straight heart tube contains all embryonic segments of the heart. This, however, is an incorrect statement, since the straight heart tube consists only of the future apical trabeculated regions of the ventricles (for review, see [2]).

During the fourth and fifth week of development, the tubular heart elongates by continuous addition of new segments from the heart-forming field to its cranial and

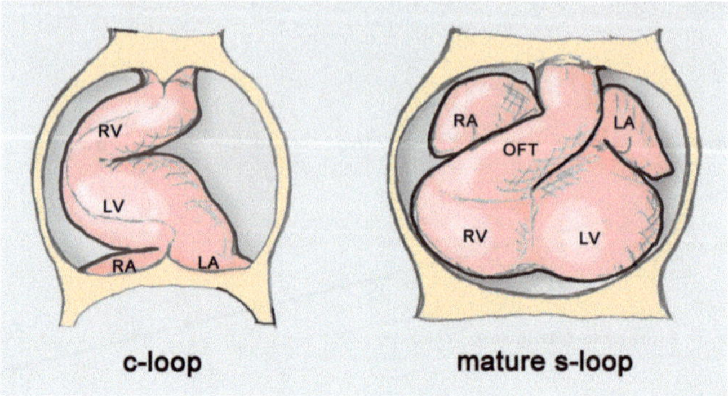

Fig. 2.2 These drawings show the normal positional changes of the embryonic ventricles caused by looping morphogenesis (ventral view into the opened pericardial cavity). (1) Caudal shift of the developing ventricles with respect to the developing atria; (2) conversion of the original craniocaudal alignment of the ventricular chambers into the definitive left–right topology of the ventricular chambers; (3) shortening of the distance between the arterial and venous heart poles. *LA* left atrium; *LV* left ventricle; *OFT* outflow tract; *RA* right atrium; *RV* right ventricle

caudal ends. Due to the limited space within the embryonic pericardial cavity, the elongating heart tube undergoes bending and torsion [3] and thereby acquires the configuration of a helically wound heart loop [4]. This phase of cardiac morphogenesis is named *cardiac looping* as a whole, but can be subdivided into the subphases of c-looping, early s-looping and late s-looping [4]. Looping morphogenesis is said to bring the embryonic segments of the heart tube and the developing great vessels into an approximation of their definitive topographical relationships. The developing ventricles, for example, are originally aligned along the craniocaudal body axis where the so-called left ventricle has a caudal position with respect to the so-called right ventricle (Fig. 2.2). As a consequence of looping morphogenesis, this positional relationship is converted into the definitive left–right topology of the ventricles (Fig. 2.2). In view of its central role in the correct positioning of the developing heart segments, cardiac looping is regarded as the key phase in the morphogenesis of higher vertebrate hearts (for review, see [4]). Having this in mind, we find it helpful to distinguish three subsequent phases of cardiac morphogenesis (Fig. 2.1): (1) *the pre-looping phase*, which is characterized by the formation of the bilaterally paired heart-forming field and the subsequent establishment of a straight heart tube (see above); (2) *the looping phase*, which is characterized by the above-mentioned morphological and positional changes of the tubular embryonic heart; and (3) *the post-looping phase*.

The post-looping phase of cardiac morphogenesis is characterized by the transformation of the helically wound heart tube into a four-chambered heart with separate pulmonary and systemic flow pathways. This transformation is accomplished mainly by three processes: (1) ballooning, (2) septation, and (3) valvulogenesis. The term *ballooning* characterizes local growth processes leading to the outward bulging of the heart wall at circumscribed areas, which represent the developing atrial appendages and the apical trabeculated regions of the left and right ventricle [5]. The term *septation* summarizes all morphogenetic processes leading to the formation of septa, which divide the originally single cardiac flow pathway into systemic and pulmonary flow pathways. The term valvulogenesis defines all morphogenetic processes involved in the formation of the semilunar and atrioventricular valves.

By the end of the embryonic period (end of the eighth week of development), the developing heart has acquired the basic four-chambered design of the mature human heart. The stems of the coronary blood vessels and their main branches are present [6], but cardiac lymph vessels just appear at the base of the ventricles [7].

Development of the Epi- and Myocardial Blood Supply

The development of the coronary blood vessels precedes that of the cardiac lymphatic vessels, both in time and place [7, 8]. It, therefore, seems reasonable to start with a description of the development of the blood supply of the embryonic heart before we will focus on the formation of the cardiac lymphatic system.

The human embryonic heart starts its pumping action around the twenty-first day of development [9, 10]. At this time point, it is seen as a tubular peristaltic pump whose outer wall is formed by naked myocardium that is in direct contact with the pericardial fluid. The naked embryonic heart muscle is a two-layered epithelium that not only lacks an epicardial covering but, additionally, lacks coronary blood vessels. Its blood supply, therefore, comes exclusively from the heart lumen whose endocardial lining is connected with the myocardium by a layer of cell-free extracellular matrix, called the *cardiac jelly* (Fig. 2.3a). The cardiac jelly forms a relatively thick diffusion barrier between the heart lumen and the myocardium. Its presence seems to conflict with the need for an efficient blood supply of a steadily working myocardium. It has been found, however, that its presence is needed for efficient peristaltic pumping of the tubular heart [11]. The design of the early embryonic heart, therefore, seems to be optimized for the purpose of peristaltic pumping but, on the other hand, seems to neglect the need for an efficient blood supply of its contracting muscle layer.

The situation changes during the mid-phase of cardiac looping when pouch-like invaginations of the endocardium start to remove the cardiac jelly from the apical portions of the developing ventricles (Fig. 2.3b) and, thereby, reduces the thickness of the diffusion barrier between blood and myocardium [12, 13]. The latter is, for the most part, still a naked myocardium forming the outer surface of the tubular heart. It is only at the dorsal wall of the embryonic ventricles where patches of epicardial mesothelium appear in the early s-shaped heart loop (fifth week; Carnegie stage 12; [14]). These patches of primitive epicardium are derived from a primarily extracardiac population of cardiac progenitor cells called the *proepicardium* (PE). The latter structure is a cauliflower-shaped accumulation of villous or vesicular protrusions of the pericardial mesothelium covering the ventral wall of the *sinus venosus* and the *septum transversum* [15].

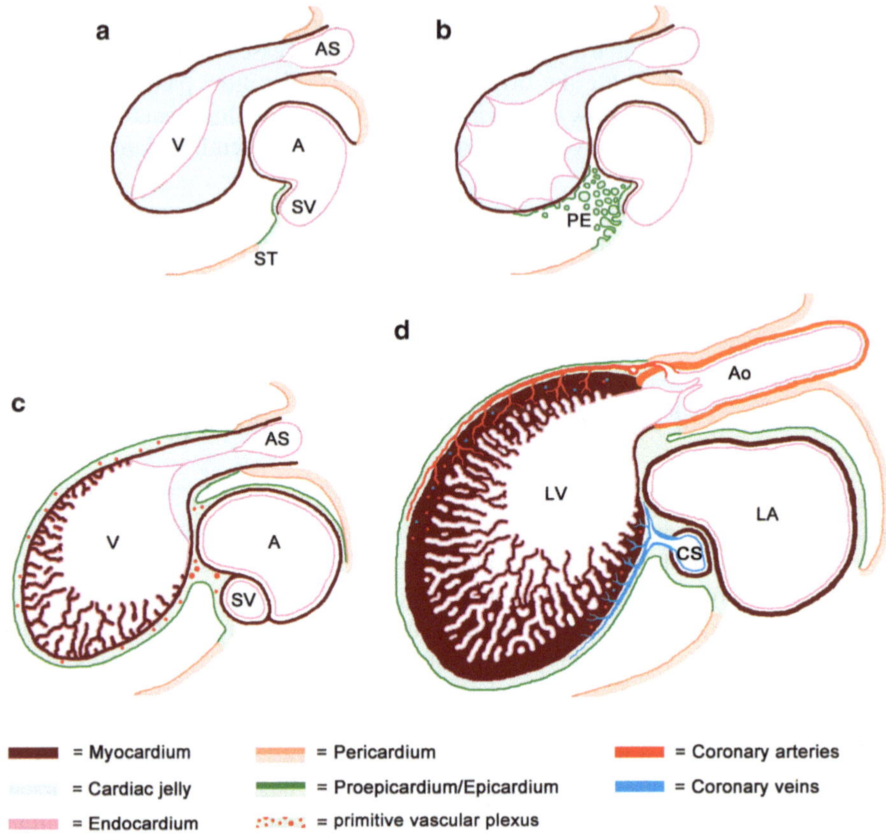

Fig. 2.3 These schematic drawings show the main steps in the development of the myocardial blood supply. (**a**) *Initial phase of blood supply from the endocardial lumen*. The wall of the early embryonic heart consists of three layers: (1) naked myocardium, (2) cardiac jelly, and (3) endocardium. The cardiac jelly is needed for efficient peristaltic pumping but forms a relatively thick diffusion barrier between the heart lumen and the myocardium. (**b**) *Transition to blood supply* via *myocardial sinusoids*. During the mid-phase of cardiac looping, pouch-like outgrowths of the endocardium start to remove the cardiac jelly from the apical portions of the developing ventricles, thereby reducing the thickness of the diffusion barrier between blood and myocardium. The myocardium is, for the most part, still a naked myocardium. Formation of the epicardium starts at the dorsal wall of the embryonic ventricles. It derives from a primarily extracardiac progenitor cell population called the proepicardium (PE). (**c**) *Blood supply* via *myocardial sinusoids*. During late stages of cardiac looping, endocardial pouches invade the inner layer of the ventricular myocardium. Centrifugal growth of the endocardial pouches and inner myocardium create the spongy/trabecular myocardium. The blood supply of the spongy myocardium comes from the endocardial pouches, which are now named the myocardial sinusoids. The outer myocardial layer is completely covered with PE-derived epicardium. Epicardium-derived cells colonize the subepicardial, myocardial and subendocardial layers of the heart. Epicardium-derived endothelial cells form a primitive vascular plexus within the subepicardial mesenchyme, which is the anlage of the coronary blood vessels. (**d**) *Blood supply* via *coronary blood vessels*. The subepicardial vascular plexus establishes open connections with the sinus venosus and the ascending aorta. Remodeling of the plexus forms a hierarchically arranged vascular system of coronary arteries, capillaries and veins. Myocardial blood supply via coronary blood vessels gradually replaces blood supply via myocardial sinusoids. *A* undivided embryonic atrium; *Ao* aorta; *AS* aortic sac; *CS* coronary sinus; *LA* left atrium; *LV* left ventricle; *PE* proepicardium; *ST* septum transversum; *SV* sinus venosus; *V* undivided embryonic ventricle

During advanced stages of cardiac looping, endocardial pouches reach the inner layer of the ventricular myocardium and start to invade the heart muscle [12]. Signals from the invading endocardial pouches stimulate proliferation of the surrounding myocardium [16]. Consequently, both the endocardial pouches and inner myocardium grow together in a centrifugal pattern, thereby, creating an inner myocardial layer of spongy architecture (Fig. 2.3c). The spongy myocardium is frequently named the *trabecular myocardium* and the process of its formation is called *trabeculation*. The spongy myocardium still lacks coronary blood vessels but appears like a highly vascularized tissue since it is in intimate contact with the blood-filled lumina of the endocardial pouches, which are now named the *myocardial sinusoids* or *intertrabecular spaces*. The blind-ending myocardial sinusoids become filled and emptied with blood during the cardiac cycle and, thereby, facilitate a relatively efficient blood supply of the avascular embryonic myocardium.

While the inner myocardial layer of the developing ventricles transforms into a blood-filled sponge, the outer myocardial layer of the human embryonic heart remains compact and becomes completely covered with epicardial mesothelium derived from the PE (fifth + sixth week; Carnegie stages 12–16; [14]). The PE and primitive epicardium play a central role in the formation of the coronary blood vessels since they provide not only the mesothelial covering of the heart but, additionally, provide mesenchymal cells that colonize the subepicardial, myocardial, and subendocardial layers where they differentiate into various cell types such as fibroblasts, coronary endothelial cells, and coronary smooth muscle cells [15]. The formation of the coronary blood vessels starts within the subepicardial mesenchyme from where it progresses into the myocardium. Epicardium-derived endothelial cells form a primitive coronary vascular plexus by vasculogenesis (sixth week; Carnegie stages 14/15; [17]). Initially, this plexus is neither connected with the *sinus venosus* nor with the developing great arterial trunks and, therefore, does not contribute to myocardial blood supply during this developmental period.

The situation changes, however, when the subepicardial part of the coronary vascular plexus establishes open connections firstly with the *sinus venosus* (sixth week; Carnegie stages 15–17) and later with the ascending aorta (end of sixth week; Carnegie stage 18; [17]). Now, blood starts to flow through the plexus, and the hemodynamic stimuli trigger the remodeling of the plexus into a hierarchically arranged vascular system of coronary arteries, capillaries, and veins (Fig. 2.3d). The remodeling process starts at the base of the ventricles from where it progresses towards the cardiac apex. As a consequence, the stems of the coronary blood vessels as well as their main branches are present at the end of the embryonic period (end of eighth week; Carnegie stage 23; [6]). Myocardial blood supply via coronary blood vessels gradually replaces blood supply via myocardial sinusoids during the late embryonic and early fetal periods. This is accompanied by expansion of the outer compact layer of the ventricular myocardium and reduction of the inner spongy layer of the ventricular myocardium.

Mature Pattern of Cardiac Lymphatics

The lymphatics of the heart were first described by Rudbeck [18] and were also noted early on by Mascagni [19] (cited from [20]). Cardiac lymphatics are formed by three interconnected parts: subendocardial, myocardial, and subepicardial plexuses. The subendocardial plexus is regularly present in the cardiac walls of the human heart, but rarely found in the atrioventricular and semilunar valves, and absent in the *Chordae tendineae* and the intima of the large arteries [21]. Lymph of the subendocardial plexus is drained by lymphatics that traverse the myocardium along branches of the coronary arteries into the subepicardial plexus [7, 22, 23]. The subepicardial plexus forms a rhomboid pattern of initial lymphatics, which, however, do not cross the interventricular sulci. Thereby a left and a right drainage territory are formed. The lymphatic capillaries discharge into paired collectors, largely following the branches of the coronary arteries. The left and the right coronary lymphatic trunks only form sparse interconnections, mainly at the base of the ascending aorta. The right coronary lymphatic trunk runs ventrally along the *aortic arch* and the *left subclavian artery* and drains into the left venous angle. The left coronary lymphatic trunk runs dorsally along the pulmonary arterial trunk, ascends to tracheobronchial lymph nodes, and finally drains into the right venous angle [7, 23].

Development of Cardiac Lymphatics

It may be a general rule of prenatal lymphangiogenesis that the development of the lymphatics starts considerably later than that of blood vessels. The earliest morphological signs of the lymphatic vascular system are the JLS, which develop at the confluence of the cranial and caudal cardinal veins. In the human, JLS were found in 6- to 7-week-old embryos of 10–14 mm total length [24], which is 3–4 weeks after the development of the first blood vessels. In the mouse, blood vascular development starts at embryonic day (ED) 7.5 [25] whereas the anlagen of the lymphatics are visible at ED 10 [26].

Corresponding to the general timing of prenatal lymphangiogenesis, it has been found that the development of the cardiac lymphatics follows, in time and space, the development of the coronary blood vessels [7, 8, 27–30]. An analysis of the developmental patterning of the cardiac parasympathetic nerves in chick embryos, as shown by Kuratani and Tanaka [31] and Verberne et al. [32], also discloses a striking spatial and temporal correlation with cardiac lymphangiogenesis, whose functional significance remains to be determined. The development of human cardiac lymphatics was studied in detail by Kampmeier [7]. At the end of the embryonic period (eighth week), he found two plexiform extensions from the mediastinal lymphatics at the base of the heart of a 30 mm embryo. The first plexus was located ventral from the left carotid artery and the left aortic arch and extended from the left

JLS to the area of the right coronary artery. The second plexus was located dorsal to the pulmonary arterial trunk and extended from the right JLS to the area of the left coronary artery. Obviously, the early anlagen represent the future efferent lymphatic conduits. In 40 mm fetuses, the lymphatics extended in the subepicardium further along the coronary arteries towards the ventricular apex. By the end of the third and beginning of the fourth prenatal month, most of the surface of the human heart was covered by lymphatics, much more densely over the ventricles than the atria. Only then, ingrowth of lymphatics into the myocardium was noticed. Corresponding observations were made in the developing heart of pigs [27], arguing for an outside-in development of lymphatics in the heart. A similar pattern of cardiac lymphangiogenesis can be observed in mice [33, 34]. Development of lymphatic valves starts in the fourth month of human prenatal development in the main channels of the cardiac lymphatics, and they are much more abundant in the region of the ventricles compared to the atria [7].

The descriptive data from Kampmeier [7] as well as recent experimental data from our labs [30] argue for a predominant origin of the cardiac lymphatics from the JLSs. We have addressed the question on the origin of the cardiac lymphatic system by studying avian embryos, which are well-suited model organisms for clarifying the embryonic origin of cell lineages and organ systems [35]. Like mammals, birds possess a completely septated four-chambered heart and the pattern of formation of the lymphatic system of the chick heart is similar to the pattern observed in the human embryonic and fetal heart [8, 28]. At incubation day (ID) 9, corresponding to late stages of human embryonic development, subepicardial lymphatics are present at the base of the heart from where they grow towards the apex (Fig. 2.4a). By ID 17, corresponding to early fetal stages of human development, they cover the whole surface of the ventricles [8]. At ID 15, lymphatics are found in the myocardium, adjacent to the branches of the coronary arteries (Fig. 2.4b).

Cardiac lymph is drained by lymphatic trunks that accompany the pulmonary arterial trunk and the aorta (Fig. 2.4c). The lymphatics are often in close association with the coronary arteries and it has been found that the development of the coronary blood vessels precedes that of the cardiac lymphatic vessels, both in time and place [7, 8]. It was, therefore, tempting to speculate that the two vascular systems of the heart may share a common origin. However, whereas the coronary blood vessels originate from the proepicardium (see above), lymphatics were not formed by quail cells, when quail PE was grafted homotopically into chick embryos, indicating a non-PE origin of cardiac lymphatics (Fig. 2.5). In these quail-chick chimeras, we only observed a single exception to this rule. At the base of the heart we found one lymphatic trunk that regularly contained quail endothelial cells (Fig. 2.5c). This implies that (1) there exists a lymphovenous anastomosis at the base of the heart, and (2) this anastomosis may develop by ingrowth of a sprout from subepicardial blood vessels—similar to the homing of the subepicardial vascular plexus into the aorta and the coronary venous sinus (see above). Lymphovenous anastomoses have been found in various organ systems. In the subepicardium of human and pig hearts they occur frequently [36], but their functional significance remains to be determined.

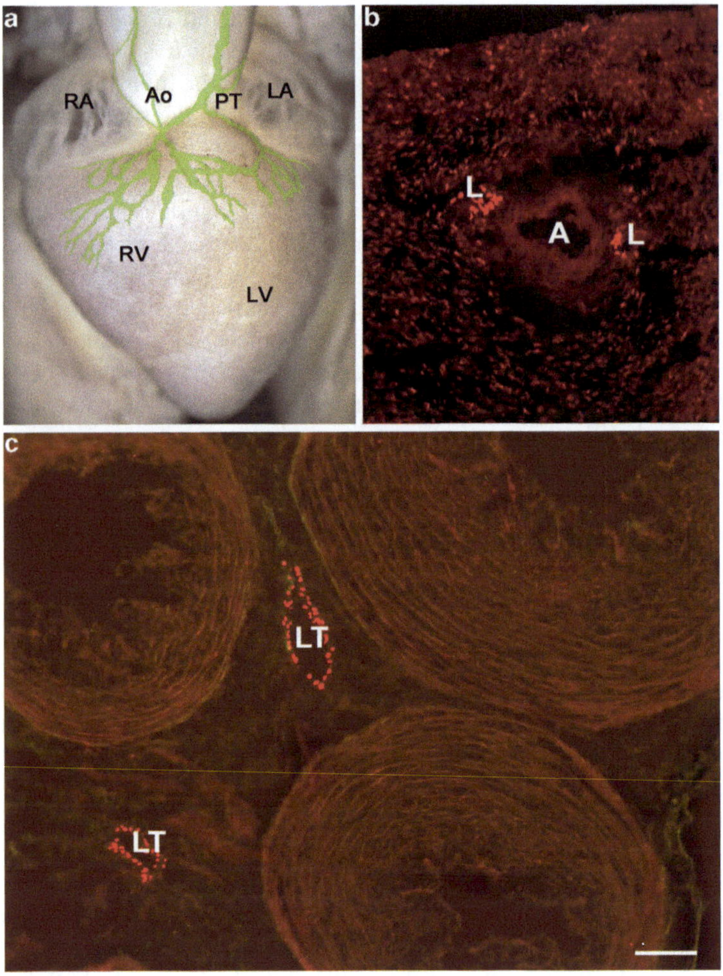

Fig. 2.4 (**a**) Chick heart at incubation day 9. Whole mount staining of subepicardial lymphatics with anti-Prox1 antibodies. For the sake of clarity, lymphatics were colorized in *green*. Lymphatics are found at the base of the heart and along the great arterial trunks. *Ao* aorta; *LA/RA* left and right atrium; *LV/RV* left and right ventricle; *PT* pulmonary trunk. (**b**) Chick heart at incubation day 15. In the myocardium Prox1-positive lymphatics (L) are found adjacent to an artery (A) (**c**) Chick heart at incubation day 15. Prox1-positive lymphatic trunks (LT) are found along the arterial outflow tracts. Bar = 100 μm

Direct connections of the developing cardiac lymphatics with the JLS were noted by Kampmeier [7], and centrifugal growth of the lymph sacs, as proposed by Sabin [37], may be the main mechanism for their development. Transdifferentiation of venous endothelium into lymphatic endothelium takes place in the jugular region, which is characterized by the confluence of the cranial and caudal cardinal veins into the common cardinal vein (Fig. 2.6). Transient expression of the transcription factor Sox18 and permanent expression of the transcription factor Prox1 are essential for the development of lymphatic endothelial cells [26, 38]. In all species studied

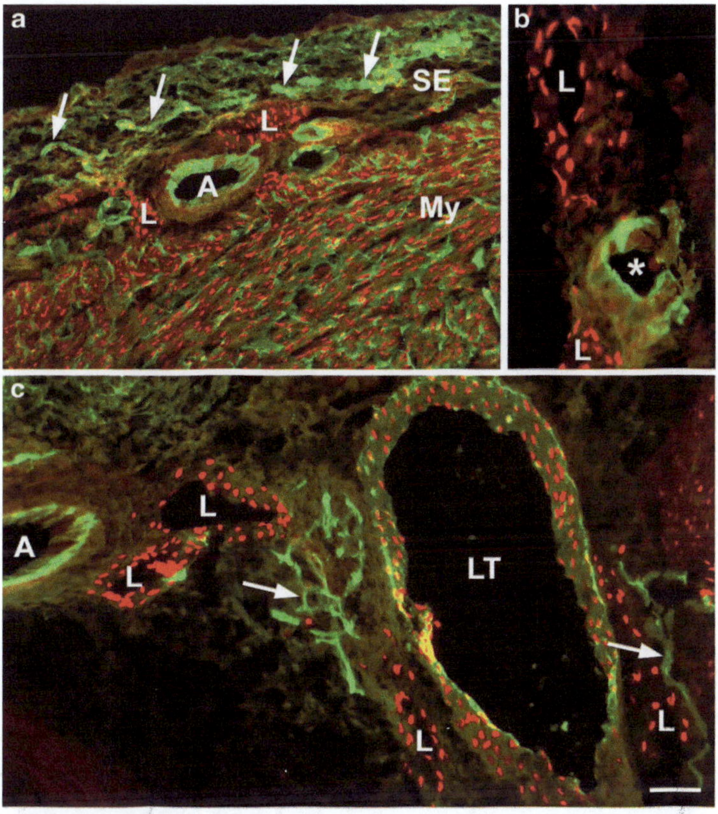

Fig. 2.5 Homotopical grafting of quail proepicardium at incubation day 2.5 into chick embryos and reincubation until incubation day 15. Staining of quail endothelial cells with QH1 antibody (*green*) and lymphatics (L) with anti-Prox1 antibody (*red*). (**a**) Blood capillaries in the myocardium (My) are of proepicardial (quail) origin. The same holds true for arteries (A), veins and capillaries (*arrows*) in the subepicardium (SE). Lymphatics (L) are not derived from the proepicardium. (**b**) Higher magnification showing QH1-positive blood vessel and QH1-negative lymphatics (L). (**c**) At the base of the heart, arteries (A) and subepicardial blood capillaries (*arrows*) are QH1-positive. Lymphatics (L) are QH1-negative, except for a lymphatic trunk (LT), which contains some quail endothelial cells. Bar = 70 μm

so far, Prox1 is expressed in a subpopulation of jugular venous endothelial cells and in the JLS (Fig. 2.7). Macrophages are in close association with the developing lymphatics (Fig. 2.7a) and may promote their growth by the secretion of the lymphangiogenic growth factor VEGF-C [39]. Cells with an intermediate macrophage-lymphendothelial phenotype have also been noted [30, 40]. Their integration into embryonic lymphatics has not been proven, but there are human pathologies where the participation of circulating cells in lymphangiogenesis is highly likely, such as Kaposi's sarcoma [41] and inflammatory kidney rejection [42]. In summary, the origin of the cardiac lymphatic system has not been clarified in detail. Fate-mapping studies on avian embryos have excluded the possibility that the cardiac lymphatics derive from the PE. The JLS are the most likely source of origin and sprouting

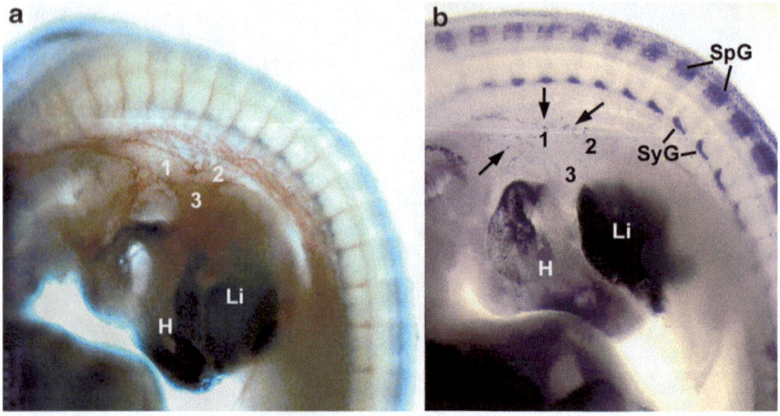

Fig. 2.6 Whole mount in situ hybridization of chick embryos at incubation day 4. (**a**) Double staining with a Tie2 probe (*red-brown*) and a Prox1 probe (*blue*). The jugular region is shown, characterized by the confluence of the cranial cardinal vein (1) with the caudal cardinal vein (2) into the common cardinal vein (3). *H* heart; *Li* liver. (**b**) Staining with a Prox1 probe reveals scattered positive cells in the jugular region (*arrows*), and signals in the spinal ganglia (SpG) as well as sympathetic ganglia (SyG). *H* heart; *Li* liver

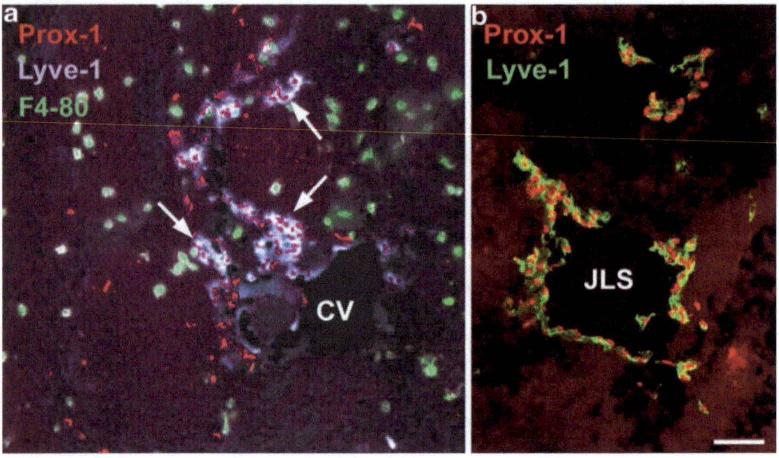

Fig. 2.7 Immunofluorescence studies of mouse embryos. (**a**) Triple staining with antibodies against Prox1 (*red*), Lyve-1 (*blue*) and F4-80 (macrophages, *green*). At embryonic day 10, Prox1 and Lyve-1 double positive lymphatic anlagen are seen, which obviously originate from the cardinal vein (CV). Some macrophages are in immediate contact with the lymphatics. (**b**) At embryonic day 11, Prox1 and Lyve-1 double positive jugular lymph sacs (JLS) have developed. Bar = 60 μm (Specimens provided by Dr. K. Buttler, Göttingen, Germany)

angiogenesis may be the dominant mechanism in cardiac lymphangiogenesis. The possible contribution of scattered mesenchymal cells to the developing cardiac lymphatics, as suggested by descriptive data [28, 43–45], needs further investigation.

Conclusions and Outlook

As compared to the coronary blood vessels, the development and function of the cardiac lymphatics are much less understood. The origin of cardiac lymphatics has not been established unequivocally, and the mechanisms of their formation (lymphangiogenesis, lymphvasculogenesis) remain to be clarified. Recent progress in the development of transgenic animals carrying reporter genes to visualize the lymphatics will bring more light into the darkness.

The functional significance of the lymphatics in the heart was only recognized by a few specialists in the field. Myocardial edema decreases cardiac output and, when chronic, causes interstitial fibrosis [46]. In general, the impairment of the cardiac lymph flow, induced by myocardial injury or various surgical procedures, was never connected with recovery and patient outcome [47]. With water-sensitive MRI the visualization of myocardial edema is possible in vivo and can be used as a diagnostic tool. In combination with scar imaging it differentiates between reversible and irreversible damage and provides prognostic information [48].

Furthermore, the cross-talk between the lymphatics and the adjacent cardiac cells has never been studied. This is therefore an open field for regenerative medicine.

References

1. Vincent SD, Buckingham ME (2010) How to make a heart: the origin and regulation of cardiac progenitor cells. Curr Top Dev Biol 90:1–41
2. Van den Berg G, Moorman AFM (2009) Concepts of cardiac development in retrospect. Pediatr Cardiol 30:580–587
3. Davis CL (1927) The development of the human heart from its first appearance to the stage found in embryos of twenty paired somites. Contrib Embryol 19:245–284
4. Männer J (2009) The anatomy of cardiac looping: a step towards the understanding of the morphogenesis of several forms of congenital cardiac malformations. Clin Anat 22:21–35
5. Moorman A, Webb S, Brown NA, Lamers W, Anderson RH (2003) Development of the heart: (1) formation of the cardiac chambers and arterial trunks. Heart 89:806–814
6. Licata JM (1954) The human embryonic heart in the ninth week. Am J Anat 94:73–125
7. Kampmeier OF (1928) On the lymph flow of the human heart, with reference to the development of the channels and the first appearance, distribution, and physiology of their valves. Am Heart J 4:210–222
8. Rychter Z, Jelinek R, Klika E, Antalikova L (1971) Development of the lymph bed in the wall of the chick embryo heart. Physiol Bohemoslov 20:533–539
9. Britten S, Soenksen DM, Bustillo M, Coulam CB (1994) Very early (24–56 days from the last menstrual period) embryonic heart rate in normal pregnancies. Hum Reprod 9:2424–2426
10. Wisser J, Dirschedl P (1994) Embryonic heart rate in dated human embryos. Early Hum Dev 37:107–115
11. Männer J, Thrane L, Norozi K, Yelbuz TM (2008) High resolution in vivo imaging of the cross-sectional deformations of contracting embryonic heart loops using optical coherence tomography. Dev Dyn 237:953–961

12. Icardo JM, Fernandez-Terán A (1987) Morphological study of ventricular trabeculation in the embryonic heart. Acta Anat 130:264–274
13. Männer J, Thrane L, Norozi K, Yelbuz TM (2009) In vivo imaging of the cyclic changes in cross-sectional shape of the ventricular segment of pulsating embryonic chick hearts at stages 14 to 17: a contribution to the understanding of the ontogenesis of cardiac pumping function. Dev Dyn 238:3273–3284
14. Hirakow R (1989) Epicardial formation in staged human embryos. Acta Anat Nippon 67:616–622
15. Männer J, Pérez-Pomarez JM, Macías D, Muñoz-Chápuli R (2001) The origin, formation, and developmental significance of the epicardium: a review. Cells Tissues Organs 169:89–103
16. Peshkovsky C, Totong R, Yelon D (2011) Dependence of cardiac trabeculation on neuregulin signaling and blood flow in zebrafish. Dev Dyn 240:446–456
17. Hutchins GM, Kessler-Hanna A, Moore GW (1988) Development of the coronary arteries in the embryonic human heart. Circulation 77:1250–1257
18. Rudbeck O (1653) Nova exercitatio anatomica exhibens Ductus Hepaticos Aquosos et Vasa Glandularum Serosa, nunc primum in venta, aeneisque figuris delineata. Arosia
19. Mascagni P (1787) Vasorum Lymphaticorum Corporis Humani Descriptio et Iconographia. Pazzini Carli, Siena, Italy
20. Rusznyak I, Földi M, Szabo G (1969) Lymphologie: Physiologie und Pathologie der Lymphgefässe und des Lymphkreislaufes. G. Fischer, Stuttgart, Germany, p 625
21. Eberth CJ, Belajeff A (1886) Über die lymphgefässe des Herzens. Arch Pathol Anat 37:124
22. Sacchi G, Weber E, Aglianò M, Cavina N, Comparini L (1999) Lymphatic vessels of the human heart: precollectors and collecting vessels: a morpho-structural study. J Submicrosc Cytol Pathol 31:515–525
23. Kubik S (2003) Anatomy of the lymphatic system. In: Földi M, Földi E, Kubik S (eds) Textbook of lymphology. Urban und Fischer Verlag/Elsevier GmbH, München, Germany, pp 2–166
24. Van der Putte SCJ (1975) The development of the lymphatic system in man. Adv Anat Embryol Cell Biol 51:1–60
25. Breier G, Breviario F, Caveda L et al (1966) Molecular cloning and expression of murine vascular endothelial-cadherin in early stage development of cardiovascular system. Blood 87:630–641
26. Wigle JT, Oliver G (1999) Prox1 function is required for the development of the murine lymphatic system. Cell 98:769–778
27. Cash JR (1917) On the development of the lymphatics in the heart of the embryo pig. Anat Rec 13:451–464
28. Klika E, Antalikova L, Rychter Z, Jelinek R (1972) Inception and manner of development of the lymph vessels in the chick embryo heart. Lymphology 5:137–148
29. Riquet M, Briere J, Dupont P, Pennhouat G, Hidden G (1993) The embryonic and early fetal development of the lymphatics of the heart and lungs in humans. Surg Radiol Anat 15:369–370
30. Wilting J, Buttler K, Schulte I, Papoutsi M, Schweigerer L, Männer J (2007) The proepicardium of avian embryo delivers hemangioblasts but not lymphangioblasts to the developing heart. Dev Biol 305:451–459
31. Kuratani S, Tanaka S (1990) Peripheral development of the avian vagus nerve with special reference to the morphological innervation of heart and lung. Anat Embryol 182:435–445
32. Verberne ME, Gittenberger-de Groot AC, Poelmann RE (1998) Lineage and development of the parasympathetic nervous system of the embryonic chick heart. Anat Embryol 198:171–184
33. Mäkinen T, Jussila L, Veikkola T, Karpanen T, Kettunen MI, Pulkkanen KJ et al (2001) Inhibition of lymphangiogenesis with resulting lymphedema in transgenic mice expressing soluble VEGF receptor-3. Nat Med 7:199–205

34. Xu Y, Yuan L, Mak J, Pardanaud L, Caunt M, Kasman I et al (2010) Neuropilin-2 mediates VEGF-C-induced lymphatic sprouting together with VEGFR3. J Cell Biol 188:115–130
35. Le Douarin NM (2008) Developmental patterning deciphered in avian chimeras. Dev Growth Differ 50(suppl 1):S11–S28
36. Vajda J, Tomcsik M, van Doorenmaalen WJ (1972) Connections between the venous system of the heart and the epicardiac lymphatic network. Acta Anat (Basel) 83(2):262–274
37. Sabin FR (1902) On the origin of the lymphatic system from the veins and the development of the lymph hearts and thoracic duct in the pig. Am J Anat 1:367–389
38. François M, Caprini A, Hosking B, Orsenigo F, Wilhelm D, Browne C et al (2008) Sox18 induces development of the lymphatic vasculature in mice. Nature 456:643–647
39. Schoppmann SF, Birner P, Stöckl J, Kalt R, Ullrich R, Caucig C et al (2002) Tumor-associated macrophages express lymphatic endothelial growth factors and are related to peritumoral lymphangiogenesis. Am J Pathol 161:947–956
40. Buttler K, Kreysing A, von Kaisenberg CS, Schweigerer L, Gale N, Papoutsi M et al (2006) Mesenchymal cells with leukocyte and lymphendothelial characteristics in murine embryos. Dev Dyn 235:1554–1562
41. Barozzi P, Luppi M, Facchetti F, Mecucci C, Alù M, Sarid R et al (2003) Post-transplant Kaposi sarcoma originates from the seeding of donor-derived progenitors. Nat Med 9:554–561
42. Kerjaschki D, Huttary N, Raab I, Regele H, Bojarski-Nagy K, Bartel G et al (2006) Lymphatic endothelial progenitor cells contribute to de novo lymphangiogenesis in human renal transplants. Nat Med 12:230–234
43. Juszynski M, Ciszek B, Stachurska E, Jablonska A, Ratajska A (2008) Development of lymphatic vessels in mouse embryonic and postnatal hearts. Dev Dyn 237:2973–2986
44. Karunamuni G, Yang K, Doughman YQ, Wikenheiser J, Bader D, Barnett J et al (2010) Expression of lymphatic markers during avian and mouse cardiogenesis. Anat Rec 293:259–270
45. Flaht A, Jankowska-Steifer E, Radomska DM, Madej M, Gula G, Kujawa M et al (2012) Cellular phenotypes and spatio-temporal patterns of lymphatic vessel development in embryonic mouse hearts. Dev Dyn 241:1473–1486
46. Laine GA, Allen SJ (1991) Left ventricular myocardial edema. Lymph flow, interstitial fibrosis, and cardiac function. Circ Res 68:1713–1721
47. Cui Y (2010) Impact of lymphatic vessels on the heart. Thorac Cardiovasc Surg 58:1–7
48. Friedrich MG (2010) Myocardial edema: a new clinical entity? Nat Rev Cardiol 7:292–296

Part II
The Cardiac Lymphatics and Heart Disease

Chapter 3
The Link Between Lymphatic Obstruction and Congenital Heart Disease

Manish Bansal

Abstract There is a complex interplay between congenital heart disease and lymphatic obstruction as evidenced by the literature where associations have been reported between lymphatic impairment and congenital heart disease. Turner syndrome is a classic example of a condition where lymphatic impairment might be the cause of some congenital heart diseases. At the same time there are congenital heart diseases which impair the lymphatic flow, leading to significant morbidity and mortality; these include left-to-right shunt lesions and protein-losing enteropathy. Congenital cardiac surgery which involves extensive mediastinal dissection can also cause injury to the lymphatic system. Hence, the lymphatic system must be given importance while managing patients with congenital heart disease.

Keywords Congenital heart disease • Lymphatic obstruction • Turner syndrome • Cardiac surgery • Protein-losing enteropathy

Introduction

The lymphatic system maintains homeostasis by receiving proteins and excess fluid from the interstitial spaces and returning them to the venous system [1]. Large lymphatic channels travel with the major coronary arteries in the epicardium and small lymphatics can be found in the endocardium [2]. Impairment of cardiac lymphatic flow has been known to be associated with ventricular fibrillation, increased

M. Bansal (✉)
Department of Pediatrics/Pediatric Cardiology, UH Rainbow Babies and Children's Hospital, Case Western Reserve University, 11100 Euclid Avenue, Cleveland, OH 44106, USA
e-mail: drmanishbansal@gmail.com

G. Karunamuni (ed.), *The Cardiac Lymphatic System: An Overview*,
DOI 10.1007/978-1-4614-6774-8_3,

superior vena cava pressures, and pulmonary arterial hypertension [3]. Many congenital heart surgeries involve excision or destruction of the intrathoracic and mediastinal lymphatics, primarily because it is assumed that the mediastinal lymphatic system is surgically expendable. However, obstruction of the cardiac lymphatic obstruction may lead to cardiac dysfunction and cardiac lymph edema [4]. Similarly, pulmonary lymphatic obstruction can cause pulmonary perivascular lymphedema, endothelial injury, and pulmonary arterial obstruction. In this chapter we will address the issue of cardiac lymphatic obstruction and its impact on congenital heart disease and vice versa.

Link Between Lymphatic Obstruction and Congenital Heart Disease

Increased Nuchal Translucency and Congenital Heart Disease

Increased nuchal translucency (NT) has been strongly associated with congenital heart disease [5]. This finding, as often happens, was a by-product of other studies that were primarily concerned with screening for chromosomal abnormalities. While risks for chromosomal abnormalities are adjusted for maternal age and serum biochemistry, risks for congenital heart disease appear to be solely dependent on the degree of NT itself.

Initially increased NT was thought to be associated with chromosomal abnormalities. Later it was found out that in euploid fetuses with increased NT, there is increased incidence of congenital heart diseases, the most common being the narrowing of the aortic isthmus [6]. The incidence of congenital heart disease in fetuses with increased NT and normal karyotype varies with the degree of NT, and approximately one third of fetuses with major congenital heart disease can potentially be identified by NT screening. Increased NT however does not predict the type of cardiac abnormality that may be encountered [5].

The etiology of increased NT has been widely debated with etiologies including cardiac heart disease itself, cardiac failure, and lymphatic system abnormalities. A mesenchyme-lined fluid-filled cavity (edema) was found in the posterior nuchal region together with a bilaterally enlarged jugular lymphatic system (JLS) in mutant mouse models (trisomy 16, equivalent to human trisomy 21) (Fig. 3.1) [7]. The persistent JLS were also seen by ultrasound in a large proportion of human fetuses with increased nuchal thickness. A possible delay in the development of the lymphatic vessels in the neck has been suggested to cause increased NT [8, 9]. The JLS is also the first part of the lymphatic system to develop, and a delay in such development of these vessels would lead to fluid accumulation in the neck region. As the process is only delayed, the fluid is eventually drained away when the JLS is finally able to reconnect to the venous system.

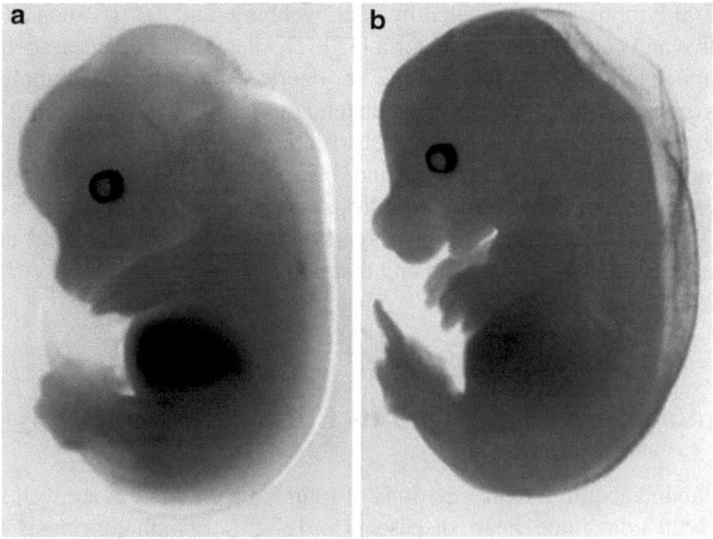

Fig. 3.1 Embryonic mouse. (**a**) Wild-type mouse embryo, day 14 of development. (**b**) Trisomy 16 embryo, day 14 of development with increased NT (Reprinted from Haak et al. [7] and used with permission of Elsevier)

Lymphatic Obstruction and Chromosomal Abnormalities

Disruption of the normal lymphatic system has been seen in patients with congenital heart diseases associated with various syndromes such as Turner and Noonan [10–12]. There is increased incidence of congenital heart disease in individuals with Turner syndrome, as well as neck webbing [13], which is formed from the postnatal residua of nuchal cystic hygromas caused by obstructed jugular lymphatics in utero. On the basis of this observation, Clark [13] proposed that centrally localized distended lymphatics compress the developing aortic root, resulting in specific left-sided defects, including hypoplastic left heart, bicuspid aortic valve, and coarctation of the aorta as a result of low flow. There are also specific right-sided defects that can occur, such as persistent left superior vena cava, anomalous pulmonary venous return, and dilated right atrium as a result of back pressure in response to obstructions in forward flow. This view was supported by further epidemiologic observations in a study of 120 infants with neck webbing reported in the Iowa Birth Defects registry, among which 66% were found to have flow-related defects [14]. These observations came from pathology studies which were focused on the most severely affected fetuses, raising the possibility that the association between congenital heart diseases and neck webbing simply reflects the most severe phenotype in 45, X individuals rather than a specific connection between these two phenotypic features of X-chromosome deletion [15].

However, in an observational study of Turner syndrome patients who were not selected for cardiovascular disease, a significant association was made between central fetal lymphedema, signaled by neck webbing, and defects such as bicuspid aortic valve and coarctation of the aorta [15]. The anatomic defects associated with fetal lymphedema in Turner syndrome are decreased numbers of lymphatics and dilated lymphatic channels that end in distended sacs, which lack connections with the venous system. Severe lymphatic obstruction early in fetal development may cause heart failure from compression and/or impaired filling of developing cardiovascular structures, leading to fetal hydrops and demise [16]. Similarly, de Mooij et al. [17] demonstrated that Noonan syndrome fetuses of gestational age 16+0 weeks demonstrated nuchal edema and distended JLS with less tissue compared to the control fetuses.

Congenital Heart Disease and Its Effect on Lymphatic System

Congenital and acquired malformations of lymphatic circulation are well known in patients with congenital heart disease [15, 18–20]. Lymphangiectasis has been observed in infants and children with obstructive left-sided lesions, such as hypoplastic left heart syndrome with restrictive atrial septum [21, 22] or total anomalous pulmonary venous return [23]. Patients with lymphatic hypoplasia usually present with lymphoedema. Congenital heart disease is rare in these patients (a 1–4% incidence was described in one series [18]) and no particular cardiac lesion predominates. Similarly, lymphatic hyperplasia was not associated with any particular congenital heart disease, despite the incidence of 9.7% being greater than expected in the general population (0.9%) [19].

Congenital Heart Diseases Associated with Increased Pulmonary Blood Flow

Congenital heart diseases with increased pulmonary blood flow mainly include ventricular septal defects, atrial septal defects, atrioventricular canal defects, and patent ductus arteriosus. These are the most common types of congenital heart diseases. These patients often have significant morbidity which can be attributed to increased lung water, impairment of normal respiratory function, and increased metabolic burden on an already compromised cardiovascular system.

Increased pulmonary blood flow leads to increased capillary filtration of protein-poor fluid into the interstitial space and increased clearance of lymphatic fluid [24]. Reddy et al. [25] created an ovine model of chronically increased pulmonary blood flow by placing a large vascular graft between the aorta and pulmonary artery of a fetal lamb. Following spontaneous delivery, these lambs had increased pulmonary blood flow and demonstrated hemodynamic and morphologic features that mimicked the human disease. Acute increase in pulmonary blood flow can result in alterations in pulmonary vascular endothelial function including disruption in endothelium-dependent nitric oxide (NO) signaling [26].

Chronic increase in pulmonary blood flow leads to lymphatic alterations, including endothelial dysfunction, resulting in decreased lymphatic flow. There is decrease in protein-poor lymph flow which is less than expected for the increased pulmonary blood flow in such lesions [26]. These changes were seen in the ovine model, and they mimic the symptomatology of tachypnea and the failure to thrive seen in children with left-to-right shunt lesions. There is an increase in pulmonary capillary hydrostatic pressure in acute and chronic shunt due to an increase in pulmonary blood flow [26]. There is a decrease in lymphatic nitric oxide production, which could play a role in the perturbation of lymphatic function as well as the postnatal development of lymphatic network [27].

Lymphatic Obstruction and Protein-Losing Enteropathy

Protein-losing enteropathy (PLE) is a relatively uncommon complication of surgical procedures used for palliation of complex congenital heart disease. The relevant lymphatic circulation converges variably, but predictably, upon a discrete location in the central venous system (Fig. 3.2) [28, 29]. Hence, obstruction of the lymphatic system could be considered as one of the etiologies of PLE.

There have been studies demonstrating a link between lymphatic obstruction and congenital heart disease, especially in cases of PLE in patients undergoing Fontan operation [30]. In their retrospective case control study, Meadows et al. [30] found a relatively high prevalence (25%) of lymphatic disruption or central venous obstruction at the site of usual lymphatic drainage in patients with PLE when compared to controls (3%). Lymphatic obstruction was evident by MRI, angiography, or documented surgical thoracic duct ligation. Central venous catheter was not shown to be associated with PLE. This suggested that physical lymphatic obstruction may play an important, and previously unrecognized, role in the development of PLE in patients with complex congenital heart disease.

The lymphatic system in patients with Fontan physiology operates at, or near, its physiologic limit [31]. Elevated central venous pressure coherent to passive circulation in Fontan patients is transmitted to the hepatic and intestinal venous circulation, leading to increased lymph production [32]. At the same time, increased central venous pressure decreased the lymphatic return to central circulation [33, 34]. As a result, the lymphatic system operated at or near its physiological limit. There is subclinical enteric protein loss with rare decompensation to clinical PLE during unpredictable physiologic insults [35].

Lymphatic Obstruction and Left Heart Lesions

Hypoplastic left heart syndrome is one of the most extensively studied lesions, in relation to its effects on the lymphatic system. Data concerning other obstructive lesions is relatively sparse. Luciani et al. [36], in a report of a single patient with congenital pulmonary lymphagiectasis and hypoplastic left heart syndrome with a

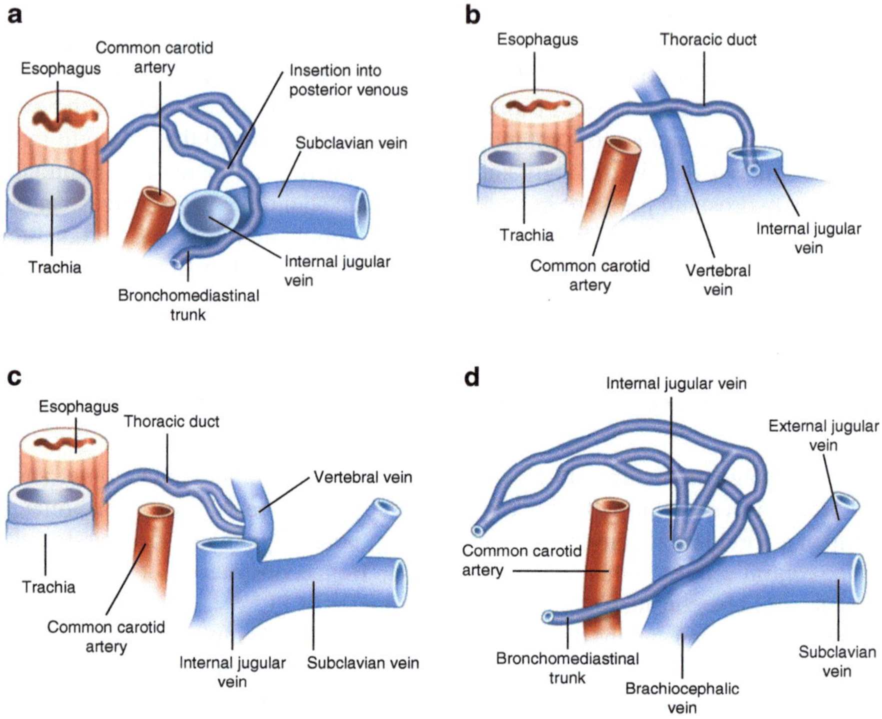

Fig. 3.2 Variations in termination of the thoracic duct. (**a**) Preterminal branching of thoracic duct and opening near the internal jugular vein. (**b**) A typical example of termination of thoracic duct into the internal jugular vein. (**c**) A bifid termination into the vertebral vein. (**d**) A complicated trifurcated termination draining into the internal jugular vein, subclavian vein, and lateral venous angle. The bronchomediastinal duct is also demonstrated (Redrawn from Langford et al. [28] and Reprinted with permission from Elsevier)

restrictive atrial septal defect, described the most severe spectrum of the lymphatic abnormalities. Graziono et al. [21] showed that four of the five patients with a restrictive ASD and hypoplastic left heart syndrome demonstrated moderate lymphatic dilatation and 1 patient had severe dilatation. In contrast, four of the five patients with nonrestrictive defects had normally lymphatics and one patient had mildly dilated lymphatics. The physiological explanation involves increased left atrial pressure in the fetal life which transmits to the pulmonary veins and the lymphatics, thus leading to the changes described above.

Cardiac Surgeries and Lymphatic Obstruction

Patients undergoing cardiac surgery often have morbidity and mortality related to pulmonary edema, impairment of normal respiratory function, and increased metabolic demands. Chylothorax is an additional cause of morbidity in patient

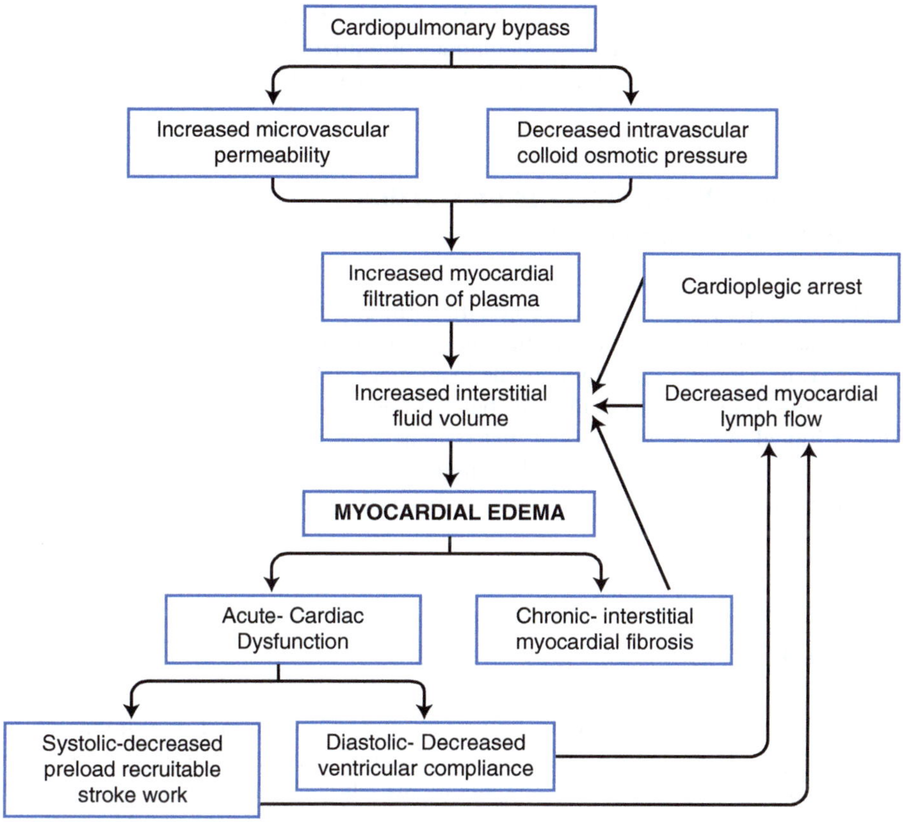

Fig. 3.3 Role of myocardial edema after cardiopulmonary bypass (Adapted from Nakamura and Rockson [38] by permission of Oxford University Press)

undergoing cardiac surgery [30]. It may be caused either by injury of the thoracic duct, increased pressure in the systemic veins exceeding that in the thoracic duct, or a central vein thrombosis.

Mehlhorn et al. [37] found that myocardial contraction is the major determinant of myocardial lymph flow and that impairment of such flow during cardioplegic arrest may contribute to postoperative myocardial edema and left ventricular dysfunction. As depicted in Fig. 3.3, cardiopulmonary bypass in children undergoing surgery for congenital heart disease leads to increased microvascular permeability and decreased intravascular colloid osmotic pressure. This results in increased myocardial filtration of the plasma thereby leading to myocardial edema. Myocardial edema thus causes systolic and diastolic dysfunction, further decreasing myocardial lymph flow and resulting in more myocardial edema [38]. As reliable means to detect myocardial edema in the clinical setting are not readily available at the bedside, many clinicians do not include this entity in their differential diagnosis of cardiac dysfunction. Knowledge of the factors involved in myocardial fluid homeostasis may help to develop techniques minimizing myocardial edema formation and may lead to better therapeutic interventions [39].

Conduction Disturbances from Cardiac Lymph Flow Impairment

Anatomical studies have demonstrated the intimate relationship between lymphatic vessels and conduction structures in which a single longitudinal lymphatic pathway always runs parallel to the right bundle in animal studies. This suggests that cardiac lymph flow impairment could contribute to conduction disturbances and arrhythmia. Clinical studies have shown that lymphedema is associated with arrhythmia [40]. It has also been proposed that higher resistance in mediastinal lymphatics could be the cause of supraventricular tachycardia as these lymphatics are responsible for draining the atria.

Summary

In conclusion, the cardiac lymphatic system plays an important role in the pathophysiology and management of patients with congenital heart disease. Also there are significant lymphatic alterations in patients who have congenital heart disease, including left-to-right shunt lesions and left heart obstructive lesions. More efforts and research should be designed to utilize this knowledge of the cardiac lymphatic system for better healthcare management of children with congenital heart disease.

References

1. Casley-Smith JR (1968) How the lymphatic system works. Lymphology 1(3):77–80
2. Loukas M, Abel N, Tubbs RS, Grabska J, Birungi J, Anderson RH (2011) The cardiac lymphatic system. Clin Anat 24(6):684–691. doi:10.1002/ca.21104; 10.1002/ca.21104
3. Lupinski RW (2009) Aortic fat pad and atrial fibrillation: Cardiac lymphatics revisited. ANZ J Surg 79(1–2):70–74. doi:10.1111/j.1445-2197.2008.04802.x
4. Cui Y, Urschel JD, Petrelli NJ (2001) The effect of cardiopulmonary lymphatic obstruction on heart and lung function. Thorac Cardiovasc Surg 49(1):35–40. doi:10.1055/s-2001-9917
5. Carvalho JS (2005) The fetal heart or the lymphatic system or …? the quest for the etiology of increased nuchal translucency. Ultrasound Obstet Gynecol 25(3):215–220. doi:10.1002/uog.1865
6. Hyett J, Moscoso G, Papapanagiotou G, Perdu M, Nicolaides KH (1996) Abnormalities of the heart and great arteries in chromosomally normal fetuses with increased nuchal translucency thickness at 11–13 weeks of gestation. Ultrasound Obstet Gynecol 7(4):245–250. doi:10.1046/j.1469-0705.1996.07040245.x

7. Haak MC, Bartelings MM, Jackson DG, Webb S, van Vugt JM, Gittenberger-de Groot AC (2002) Increased nuchal translucency is associated with jugular lymphatic distension. Hum Reprod 17(4):1086–1092
8. Bekker MN, Haak MC, Rekoert-Hollander M, Twisk J, Van Vugt JM (2005) Increased nuchal translucency and distended jugular lymphatic sacs on first-trimester ultrasound. Ultrasound Obstet Gynecol 25(3):239–245. doi:10.1002/uog.1831
9. Castelli E, Todros T, Mattutino G, Torre C, Panattoni G (2003) Light and scanning electron microscope study of nuchal translucency in a normal fetus. Ultrasound Obstet Gynecol 21(5):514–516. doi:10.1002/uog.95
10. Allanson JE (1987) Noonan syndrome. J Med Genet 24(1):9–13
11. Davenport ML (2010) Approach to the patient with turner syndrome. J Clin Endocrinol Metab 95(4):1487–1495. doi:10.1210/jc.2009-0926
12. Ferrell RE (2002) Research perspectives in inherited lymphatic disease. Ann N Y Acad Sci 979:39–51, discussion 76–9
13. Clark EB (1984) Neck web and congenital heart defects: A pathogenic association in 45 X-O turner syndrome? Teratology 29(3):355–361. doi:10.1002/tera.1420290305
14. Berdahl LD, Wenstrom KD, Hanson JW (1995) Web neck anomaly and its association with congenital heart disease. Am J Med Genet 56(3):304–307. doi:10.1002/ajmg.1320560318
15. Loscalzo ML, Van PL, Ho VB et al (2005) Association between fetal lymphedema and congenital cardiovascular defects in turner syndrome. Pediatrics 115(3):732–735. doi:10.1542/peds.2004-1369
16. Barr M Jr, Oman-Ganes L (2002) Turner syndrome morphology and morphometrics: Cardiac hypoplasia as a cause of midgestation death. Teratology 66(2):65–72. doi:10.1002/tera.10064
17. de Mooij YM, van den Akker NM, Bekker MN, Bartelings MM, van Vugt JM, Gittenberger-de Groot AC (2011) Aberrant lymphatic development in euploid fetuses with increased nuchal translucency including Noonan syndrome. Prenat Diagn 31(2):159–166. doi:10.1002/pd.2666; 10.1002/pd.2666
18. Corbett CR, Dale RF, Coltart DJ, Kinmonth JB (1982) Congenital heart disease in patients with primary lymphedemas. Lymphology 15(3):85–90
19. Dutka DP, Cousins C, Manhire AR (1991) Lymphatic abnormalities in Alagille's syndrome. Br Heart J 65(3):168–170
20. Noonan JA, Walters LR, Reeves JT (1970) Congenital pulmonary lymphangiectasis. Am J Dis Child 120(4):314–319
21. Graziano JN, Heidelberger KP, Ensing GJ, Gomez CA, Ludomirsky A (2002) The influence of a restrictive atrial septal defect on pulmonary vascular morphology in patients with hypoplastic left heart syndrome. Pediatr Cardiol 23(2):146–151. doi:10.1007/s00246-001-0038-7
22. Rychik J, Rome JJ, Collins MH, DeCampli WM, Spray TL (1999) The hypoplastic left heart syndrome with intact atrial septum: Atrial morphology, pulmonary vascular histopathology and outcome. J Am Coll Cardiol 34(2):554–560
23. Bellini C, Boccardo F, Campisi C, Bonioli E (2006) Congenital pulmonary lymphangiectasia. Orphanet J Rare Dis 1:43. doi:10.1186/1750-1172-1-43
24. Feltes TF, Hansen TN (1989) Effects of an aorticopulmonary shunt on lung fluid balance in the young lamb. Pediatr Res 26(2):94–97. doi:10.1203/00006450-198908000-00004
25. Reddy VM, Meyrick B, Wong J et al (1995) In utero placement of aortopulmonary shunts. A model of postnatal pulmonary hypertension with increased pulmonary blood flow in lambs. Circulation 92(3):606–613
26. Datar SA, Johnson EG, Oishi PE et al (2012) Altered lymphatics in an ovine model of congenital heart disease with increased pulmonary blood flow. Am J Physiol Lung Cell Mol Physiol 302(6):L530–L540. doi:10.1152/ajplung.00324.2011
27. Hagendoorn J, Padera TP, Fukumura D, Jain RK (2005) Molecular regulation of microlymphatic formation and function: Role of nitric oxide. Trends Cardiovasc Med 15(5):169–173. doi:10.1016/j.tcm.2005.06.003
28. Langford RJ, Daudia AT, Malins TJ (1999) A morphological study of the thoracic duct at the jugulo-subclavian junction. J Craniomaxillofac Surg 27(2):100–104

29. Zorzetto NL, Ripari W, De Freitas V, Seullner G (1977) Anatomical observations on the ending of the human thoracic duct. J Morphol 153(3):363–369. doi:10.1002/jmor.1051530303
30. Meadows J, Gauvreau K, Jenkins K (2008) Lymphatic obstruction and protein-losing enteropathy in patients with congenital heart disease. Congenit Heart Dis 3(4):269–276. doi:10.1111/j.1747-0803.2008.00201.x
31. Mertens L, Hagler DJ, Sauer U, Somerville J, Gewillig M (1998) Protein-losing enteropathy after the Fontan operation: An international multicenter study. PLE study group. J Thorac Cardiovasc Surg 115(5):1063–1073
32. Witte MH, Dumont AE, Clauss RH, Rader B, Levine N, Breed ES (1969) Lymph circulation in congestive heart failure: Effect of external thoracic duct drainage. Circulation 39(6):723–733
33. Szabo G, Magyar Z (1967) Effect of increased systemic venous pressure on lymph pressure and flow. Am J Physiol 212(6):1469–1474
34. Wegria R, Zekert H, Walter KE et al (1963) Effect of systemic venous pressure on drainage of lymph from thoracic duct. Am J Physiol 204:284–288
35. Lenz D, Hambsch J, Schneider P et al (2003) Protein-losing enteropathy in patients with Fontan circulation: is it triggered by infection? Crit Care 7(2):185–190
36. Luciani GB, Pessotto R, Mombello A, Mazzucco A (1999) Hypoplastic left heart syndrome with restrictive atrial septal defect and congenital pulmonary lymphangiectasis. Cardiovasc Pathol 8(1):49–51
37. Mehlhorn U, Davis KL, Burke EJ, Adams D, Laine GA, Allen SJ (1995) Impact of cardiopulmonary bypass and cardioplegic arrest on myocardial lymphatic function. Am J Physiol 268 (1 Pt 2):H178–H183
38. Nakamura K, Rockson SG (2008) The role of the lymphatic circulation in the natural history and expression of cardiovascular disease. Int J Cardiol 129(3):309–317. doi:10.1016/j.ijcard.2008.02.007
39. Mehlhorn U, Geissler HJ, Laine GA, Allen SJ (2001) Myocardial fluid balance. Eur J Cardiothorac Surg 20(6):1220–1230
40. Cui Y (2010) Impact of lymphatic vessels on the heart. Thorac Cardiovasc Surg 58(1):1–7. doi:10.1055/s-0029-1240553

Chapter 4
The Role of Lymphatics in Atherogenesis, Myocardial Infarction, Congestive Heart Failure, and Cardiac Transplantation

Shin Lin and Stanley G. Rockson

Abstract The lymphatic vasculature is a central biological participant in fluid, protein and cellular transport, and in immune responsiveness. Over the last 10 years, the biomedical investigation into the function of the lymphatic microvasculature has been vigorous, prompting reconsideration of the role of lymphatics in the genesis and progression of cardiovascular pathology. The lymphatic microvasculature of the heart and vascular wall likely participates in atherogenesis, myocardial infarction, congestive heart failure, and cardiac transplantation. Intensive exploration of lymphatic mechanisms of cardiovascular disease is likely to lead to enhanced insights and novel therapeutic approaches.

Keywords Atherosclerosis • Myocardial infarction • Congestive heart failure • Cardiac transplantation • Edema • Lymph • Lymphatics • Microvasculature

Introduction

The lymphatic vasculature plays an essential role in fluid homeostasis and in the trafficking of immunocytes [1] and is therefore critical to the edematous and immune-mediated sequelae of inflammation. In other words, the lymphatics actively participate in key structural and biological components of the inflammatory response and, thereby, represent a unique juncture for potential intervention. Active investigation into lymphatic mechanisms of disease, nevertheless, has suffered a relative lack of emphasis, due largely to an absence of suitable investigative tools and model systems [2]. Recently, powerful new lymphatic-specific markers, pharmacologic and genetic modulators, and novel investigative platforms have reinvigorated the

S. Lin • S.G. Rockson (✉)
Stanford University School of Medicine, Falk Cardiovascular Research Center, 300 Pasteur Drive, Stanford, CA 94305, USA
e-mail: rockson@stanford.edu

G. Karunamuni (ed.), *The Cardiac Lymphatic System: An Overview*,
DOI 10.1007/978-1-4614-6774-8_4, © Springer Science+Business Media New York 2013

Table 4.1 Genes involved in lymphatic development and function

Gene or gene product	Function
Angiopoietin-1	Growth factor [71]
Angiopoietin-2	Growth factor [72]
Chemokine (C-C motif) ligand 20 (CCL20)	Chemokine [73, 74]
Chemokind (C-C motif) ligand 21 (CCL21)	Chemokine [75]
Desmoplakin	Anchoring protein of intermediate filaments to the plasma membrane of adhering junctions [76]
Ephrin B2	Ligand of EphB receptors
FOXC2 (forkhead box C2)	Transcription factor [77, 78]
HGF (hepatocyte growth factor)	Growth factor [79]
Integrin α9	Adhesion molecule, possible VEGFR-3 co-receptor [80, 81]
LYVE-1	Hyaluronan receptor [58]
Macrophage mannose receptor 1	L-selectin receptor [82]
Neuropilin-2 (Nrp2)	Semaphorin and growth factor receptor [83]
Net (Elk3)	Transcription factor [84]
Plakoglobin	Connect cadherins to cytoskeleton in cell-cell junction [73, 81]
Prox1	Transcription factor [62, 85]
Podoplanin (T1α)	Transmembrane glycoprotein [86, 87]
Sex determining region Y-related high mobility group box (SOX18)	Transcription factor [88]
Syk and Src homology 2-domain containing leukocyte protein 76 (SLP-76)	Syk and SLP [89, 90]
Vascular endothelial growth factor-C (VEGF-C)	Growth factor [90, 91]
Vascular endothelial growth factor receptor-3 (VEGFR-3)	Growth factor receptor [92–94]

Source: Reprinted with permission from Nakamura K, Rockson SG [43]

study of lymphatic biology with the discovery of genes involved in lymphatic development and function (Table 4.1) [1, 3]. With this investigative renaissance, it is appropriate to reconsider lymphatic vasculature function within the context of the cardiovascular system and its complex role in the genesis, propagation, and therapeutics of cardiovascular pathology (Fig. 4.1).

The Anatomy and Function of the Myocardial Lymphatics

Anatomy of Myocardial Lymphatics

The myocardium is permeated by a dense plexus of penetrating intracardiac channels that drain interstitial fluid from the subendocardium to the subepicardium, detectable initially as lymphatic capillaries and then coalescing into collecting

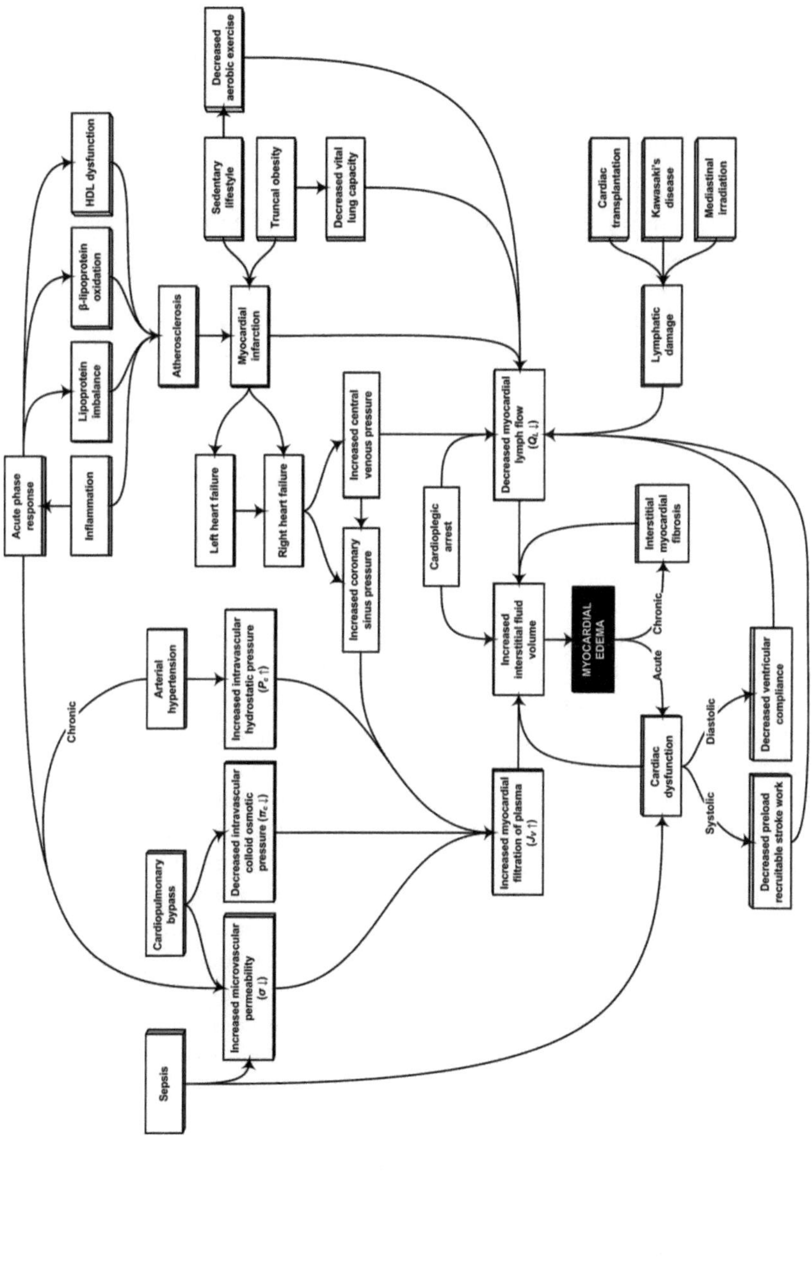

Fig. 4.1 The central role of myocardial edema in cardiac pathophysiology. A model of the theoretical causal relationships (*arrows*) in various cardiovascular pathologies as they might relate to lymphatic dysfunction and myocardial edema (Reprinted with permission from Nakamura and Rockson [43])

trunks [4]. The cardiac lymphatic trunk connects with the greater lymphatic vasculature at the cardiac lymph node before ascending to the thoracic duct and joining the central venous circulation close to the origin of the subclavian vein. In complete analogy to the peripheral lymphatic vasculature, unidirectional valves and rhythmic contraction of the adjacent tissues together ensure synchronous propulsion of myocardial lymph.

Lymphatic Function

The lymphatic system regulates interstitial fluid composition and volume both in health and in disease. The lymphatic system serves as a conduit for the by-products of normal cellular metabolism as well as those of specific pathological processes such as ischemia and necrosis [5]. As an example, inadequate perfusion of the myocardium leads to accumulation of anaerobic metabolites and disruption of myocardial fluid balance attributable to lymphatic disruption. Elucidation of these cardiac lymphatic mechanisms is expected to uncover novel strategies in the diagnosis and treatment of cardiac dysfunction.

Within a 24-h period, more than half of the circulating blood protein content is extravasated into the interstitium, yet there is no direct path of reabsorption into the arteriovenous vasculature [6]. The return to the intravascular circulation of interstitial fluid, comprised of protein, water, and other components, is the primary function of the lymphatic vasculature. The concepts of filtration and flow are central to understanding the fluid dynamics that govern lymphatic function. According to the Starling equilibrium, hydrostatic pressures from within the microvascular capillary (P_c) and circumferentially in the peripheral interstitium (P_{int}) have antagonistic effects. Intravascular hydrostatic pressure is determined largely by upstream arteriolar pressure and downstream venular pressure. In addition, colloid osmotic pressure (π) exerts an opposite effect upon directional flow, causing reabsorption of interstitial fluid back into the intravascular circulation. The final operative variable upon fluid equilibrium is the permeability of the capillary membrane. Among these physiological variables, the most important factors in fluid balance are the capillary and interstitial hydrostatic pressures. A similar arrangement of forces governs the efflux of fluid from the interstitial compartment into the lymphatic lumen. Intramyocardial fluid homeostasis is, therefore, maintained through the equilibrium that is established between fluid filtration into the myocardial interstitium and fluid flow into the lymphatic vessels. The pool of interstitial fluid exists in steady-state balance, in which disturbance of either fluid filtration or lymph flow may result in myocardial edema [7, 8]. Compensatory mechanisms help to maintain physiological conditions; of these, the most protective is the capacity to increase the rate of lymph flow (Q_{VL}) during excessive plasma filtration [4, 7]. Such augmentation is driven by increased flow of interstitial fluid into the lymphatic vessel as a consequence of increased interstitial pressure and an inverse decrease in lymphatic resistance [4, 7]. To promote lymph flow, interstitial protein concentration is also diluted in "protein washout" during higher rates of filtration [9], thereby reducing oncotic sequestration of fluid volume. Influences such as the phasic contraction of the myocardium and extracardiac thoracic movements further contribute to the dynamic regulation of fluid movement.

Anatomical and physiological study of the cardiac lymphatics in animals supports the importance of intact lymphatic function to the maintenance of tissue health [10]. Interruption of cardiac lymph circulation leads to tissue fluid stasis, inflammation, and fibrosis [10]. Acute lymphatic obstruction in the canine heart leads to epicardial lymphedema and lymphangiectasia [11]. With chronic impairment, early myocardial edema and subendocardial hemorrhage progress to endomyocardial fibrosis [10].

The Lymphatics in Cardiovascular Disease

Although recent study of the cardiac lymphatic vasculature has not been ample, the existing investigational literature suggests an important role for these vessels in the pathogenesis of atherosclerosis, myocardial infarction, congestive heart failure, and cardiac transplantation.

Atherosclerosis

Inflammation, infection, and fibrosis are the predictable consequences of lymphatic disruption in various settings of disease [12]. The presence of these events within the vascular wall may be particularly important; therefore, by inference, the loss of normal lymphatic function within the vascular wall may have a synergistic or augmenting effect upon the classically defined risk factors for atherosclerosis. In a study of human pathological specimens, it was observed that lymphatic vessels grow in areas that are rich in extracellular matrix, while regions rich in inflammatory cells are more prone to angiogenesis [13]. Furthermore, progressive atherosclerotic lesions that are rich in calcium and cholesterol crystal content demonstrate increased lymphangiogenesis in the vascular media.

Similar atherogenic mechanisms have been clinically documented following irradiation of mediastinal cardiac lymph nodes and during mediastinal lymphadenitis (which leads to severe coronary atherosclerosis in Kawasaki disease). The influence of the lymphatic system on atheroma formation may help to explain the discontinuous nature of atherosclerotic plaque formation along the axial length of the susceptible artery, as well as, potentially, the sparing of intramyocardial arteries in the face of systemically expressed risk factors.

The health of coronary arteries requires the nutritive support and metabolic equilibrium of a healthy, unimpeded circulation of the body fluids (both blood and lymph). This becomes particularly important in the context of the intramural entry and survival of apolipoprotein B-containing particles and immune cells into the arterial wall, thereby leading to the generation of pro-inflammatory and pro-atherogenic mediators. Inasmuch as the cholesterol content of atherosclerotic plaque arrives within the arterial wall through plasma filtration and is removed in lymph [14], a role for the lymphatic system in lipoprotein-mediated atherogenesis can be hypothesized. In this view, the lymphatic supply of the vascular wall itself mediates atherosclerosis through its influence upon the degree to which the arterial intima is exposed to atherogenic lipoprotein. Inadequate lymphatic flow increases the transit time of circulating lipoproteins across the arterial wall, thereby prolonging susceptibility to oxidative damage and promoting entrapment within the arterial wall. Accordingly, the lipoprotein concentration within the lymph, and presumably within the tissue of the arterial wall, is inversely related to the rate of lymph flow [15].

The cellular expression of vascular endothelial growth factors VEGF-C and VEGF-D has been reported in human monocytes and macrophages [16]. These growth factors and their cognate receptor, VEGFR-3/Flt-4, are pro-lymphangiogenic regulators expressed during various stages of development and in post-embryonic life [17]. Given the role of VEGF-C in lymphangiogenesis during wound repair, its use has been invoked therapeutically for lymphedema [12, 18]. VEGF-D signaling, interestingly, induces apoptosis of human macrophages in vivo and mononuclear cells within advanced atherosclerotic plaques [16]. The mechanistic role of this apoptosis in atherogenesis is not been completely understood. Death of lipid-laden macrophages may reduce the progression to foam cell formation and the inflammatory index of atherosclerotic lesions, while the uninterrupted phagocytosis of apoptotic debris may perpetuate inflammation and disrupt plaque stability [19].

Inflammation is a key component in the initial development of atherosclerotic lesions, but it also perpetuates disease through the promotion of plaque instability and vulnerability [20]. Both angiogenic and lymphangiogenic events are found within the inflammatory foci of plaque [21]. VEGF-C cross-activates receptors responsible for both blood and lymphatic vessel development, whereas the biological activity of VEGF-D seems to be limited to lymphangiogenesis. Despite detection of both VEGF-C and VEGF-D expression in the intima of human coronary arteries, the observable neovascularization appears to be mediated primarily through VEGF-C and through angiogenesis [21]. Differential regulation of nascent vessel formation within the atherosclerotic intima may in fact disrupt arterial-to-lymphatic vessel balance, thus creating a disequilibrium in the forces that govern fluid homeostasis. The resulting intimal edema and lymph stagnation would promote atherogenesis, as previously mentioned.

During infection, the acute-phase response provokes and potentiates the local manifestations of inflammation. These processes affect lipoprotein activity and composition; in particular, several protective proteins of HDL are functionally inactivated or displaced, rendering the immediate intimal milieu vulnerable to further oxidation and inflammation [22]. Vascular permeability is increased by the vasoactive cytokines released by activated neutrophils [23]. This promotes plasma filtration into the interstitium, thereby enhancing delivery of pro-atherogenic lipids and plasma proteins. Therefore, it is proposed that the coronary arteries become exquisitely sensitive to pro-atherogenic phenomena during the acute-phase response; paradoxically, this occurs when the lymphatics are least able to accommodate the pathological changes associated with lymph stasis [24].

In order to generate bulk lymphatic flow, the activity of the lymphatic system is predominantly modulated by the gross movements and positional changes of the thoracic cavity. Accordingly, the decreased thoracic movement and intrathoracic pressure observed in hypopnea reduces the flow of lymph, whereas aerobic exercise can increase lymph flow rates by nearly 300 % [15]. The epicardial lymphatics are especially dependent on extracardiac motion since lymph flow is impeded by the propulsive contractions of the heart, reducing their effective clearance capacity. Accordingly, the epicardial arteries are subjected to additional risk for lymph stasis and, thereby, to impaired maintenance of healthy vascular biology. Atherosclerosis is, indeed, limited nearly exclusively to the epicardial arteries [25], perhaps reflecting, at least in part, the lymphatic contribution. Common causes of sustained hypopnea, such as sedimentary lifestyle [26], decreased vital capacity, and truncal obesity [15], can thus confer independent risk for atherosclerosis explained by reduced lymphatic function. Age, hormonal status, and heredity are also implicated in the potential relationship between relative lymphatic vascular insufficiency and atherogenesis.

Myocardial Infarction

Chronic ischemia and myocardial infarction are the direct functional consequence of established and progressive atherosclerosis. When directly examined, there is a clear focal increase in the density of lymphatic vessels that is demonstrable in both acute and chronic ischemia [13]. However, this increase in lymphatic density is limited to specific pathological zones, such as necrotic edges, scars, and reactive pericarditis. Furthermore, ischemia is accompanied by neovascularization, since both blood and lymphatic vasculature demonstrate dilatation, branching, and sprouting.

Several lines of evidence support the pathophysiologic role of altered myocardial lymph flow, studied largely in canine models of myocardial infarction (MI). Experimental obstruction of cardiac lymph drainage, without compensating cessation of interstitial fluid filtration, invariably produces myocardial edema within hours [27]. Immediately following an acute coronary artery occlusion, there is a decrease of fluid efflux into the interstitium, yet lymph flow increases dramatically within the first 30 min. This phenomenon likely reflects the impact of many factors, including partial recovery of myocardial function and collateralization of myocardial perfusion. Venoconstriction occurs in response to sympathetic activation, further augmenting intracapillary pressures and, as a consequence, plasma filtration. Additionally, ischemic injury to the capillary endothelium increases permeability to plasma, augmenting both plasma filtration rates and ultrafiltrate concentrations.

The interstitial content of protein and blood products progressively rises, while the pH of the myocardial lymph falls in proportion to increasing lactate concentrations. Within the first hours of ischemia, enzyme concentrations, including lactate dehydrogenase, serum glutamic oxaloacetic transaminase, and creatine kinase, are preferentially elevated in cardiac lymph when compared with venous serum. Concomitant increases in lymph flow elevate the fraction of extracellular fluid volume occupied by lymph, ensuring that these enzymatic changes are pronounced. In MI, release of creatine kinase from the heart correlates with the degree of myocardial necrosis but may be affected by variable transport and inactivation by lymph, thus complicating the use of these biomarkers for severity and prognosis.

In aggregate, lymph flow augmentation of >50 % is observed during experimentally induced MI. Such increases, however, cannot forestall the development of persistent edema in the interstitial and vascular spaces. When edema formation occurs within the interstitium of the freshly infarcted heart, structural and functional remodeling of the myocardium occurs, particularly in the ventricular endomyocardium where the metabolic demands are highest [5]. In canine models of myocardial interstitial edema, diminution of cardiac output of up to 40 % is observed for any given level of preload, demonstrating the profound functional consequence of extravascular fluid accumulation in the myocardium [8].

Within hours of lymphatic obstruction, acute structural alterations will include myofibril degeneration, subendocardial edema, and hemorrhage [11]. Fluid accumulation itself represents a restrictive loss of compliance and cardiac function [28]. Expansion of the interstitial fluid compartment increases the diffusion distance for oxygen and exacerbates the hypoxic state, increasing the rate and magnitude of infarct development [29]. The severity of the congestion induced by experimental ligation of the major cardiac lymphatic trunks in dogs is such that coronary capillaries are compressed [11], which exacerbates the generalized hypoxia of lymph stasis. In the chronic state, this directly provokes coronary arteriopathy, with subendothelial edema and degeneration of smooth muscle with fibrinoid necrosis [30]. In murine models of ischemic injury with subsequent obstruction of lymphatic flow, myocardial and cardiac vascular fibrosis is not uncommon, compromising cardiac output and compounding the ischemic damage caused by the antecedent anoxia [31]. These experimental findings were recently corroborated by histopathological study of human autopsy specimens [32]. Although the precise mechanism through which chronic myocardial edema promotes fibrosis remains poorly understood, it is conceivable that the pathophysiology mirrors the architectural changes observed in chronic, peripheral lymphatic vascular insufficiency [33] for which tissue inflammation is a hallmark [12].

Primary collagen accumulation is a plausible mechanism for the development of myocardial fibrosis [8]. This hypothesis is supported by recent evidence demonstrating the synthesis and deposition of collagen I and III within interstitial tissues following disruption of cardiac lymph flow in rabbits [25, 34]. Within 2 days of the onset of lymph stasis, lymphatic vessels become dilated and acute inflammatory cells infiltrate the perivascular tissues and release pro-inflammatory cytokines that ultimately cause fibrosis [30]. Arterial and lymphatic metabolism shifts towards anaerobic glycolysis. These changes are most prominent in the most vulnerable vessels, including those with small luminal diameter or low reciprocity. Myocardial edema is, therefore, further exacerbated as the transport capacity of the lymphatics is overwhelmed. The accumulation of toxic by-products leads to lymphatic endothelial dysfunction and destruction and, ultimately, to complete decompensation of the lymphatic system.

Reperfusion of ischemic myocardial tissue with hyperosmolar fluid ameliorates edema with a resultant reduction in infarct size [5, 28]. Similarly, treatment of myocardial infarction with hyaluronidase, a well-recognized historical lymphagogue, produced salutary results [35–37] in animal models and in early clinical trials. Hyaluronidase infusion during experimental ischemia/reperfusion injury significantly increases cardiac lymph flow, alleviating myocardial edema and accelerating functional recovery following reperfusion; this result is independent of any appreciable increase in coronary collateralization or blood flow [37]. Furthermore, several randomized controlled trials have demonstrated mortality benefit from hyaluronidase-based pharmacotherapy of myocardial infarction [38]. Nevertheless, the benefit was modest, at best, and required treatment within 6 h of chest pain onset, limiting widespread clinical applicability [38–41].

Reperfusion alone can restore lymphatic drainage capacity to physiologic levels [5], emphasizing the clinical imperative to restore coronary patency. In the interim, adjunctive therapy to revascularization may hasten edema resolution, particularly in situations of irreparable tissue necrosis and functional deficit. Acute MI represents a dynamic complex of multiple processes and a variety of potential therapeutic targets. Augmentation of lymphatic clearance by hyaluronidase represents one out of several evidence-based interventions that improve clinical outcome. Hyaluronidase depolymerizes specific acid mucopolysaccharides and reduces inflammatory exudates within the interstitium, thereby reducing resistance and improving both interstitial fluid and coronary blood flow [42]. During the evolution of MI, hyaluronidase facilitates recovery of homeostatic blood and lymph exchange [42] to attenuate hypoxia and ATP depletion, to limit reduced myocardial and cardiac lymph flow, and to minimize toxic metabolite accumulation. Hyaluronidase thus reduces the vulnerability of the myocardium to ischemic injury by increasing cardiac lymph flow [28]. Furthermore, the increased fluid flux through the interstitial space dilutes and clears the toxic metabolites that mediate reperfusion injury. This augmented fluid filtration during reperfusion is well tolerated and does not promote further edema formation, in view of corresponding increases in downstream lymph exchange [4]. As previously discussed, while the capacity of this compensatory mechanism is lost during ischemic insult, it can be restored following reperfusion.

Immunohistochemical analysis of the known markers of lymphatic vasculature suggests that there is increased lymphangiogenesis in ischemic hearts, both acutely and chronically [43]. The lymphatic neovasculature is most prominent in the epicardium. Of considerable interest as well is the evidence that suggests increased lymphangiogenesis in atherosclerotic lesions [13].

Heart Failure

Perturbation of myocardial fluid homeostasis will produce several well-documented consequences in both systolic and diastolic function [4, 8]. Preload-recruitable stroke work is directly correlated to the extent of myocardial edema in numerous experimental settings [4]. Decreases in inherent myocardial contractility translate into decreased cardiac output, establishing myocardial edema as an independent cause of functional cardiac impairment in systole. Lymph flow rates are reciprocally dependent upon the cardiac contractile capacity. In addition, administration of a positive ionotrope enhances myocardial lymphatic function in canine models [44].

Reduction of diastolic function is thought to be a consequence of decreased ventricular compliance. Interstitial fluid accumulation, for example, can reduce the potential for myocardial expansion and therefore decrease passive ventricular filling. The edematous myocardium is further stressed by increased metabolic demands. With each systole, the edematous heart must accommodate not only decreased lymph flow but also the added viscosity of excess interstitial fluid. The anatomical and histological architecture of the heart may also become deformed, further affecting myocardial efficiency [8]. The diffusion distance also increases with edema accumulation, as myocytes are displaced farther from the capillary delivery of oxygen. Hypoxic injury is typified by anaerobic evolution of toxic metabolites, decreased cardiac contractility, and increased microvascular permeability to proteins, thereby increasing interstitial colloid pressure and fluid accumulation [11]. Chronic edema induces fibrotic changes within the interstitium of the heart [8], as does edema secondary to insults such as hypoxic injury. There is interstitial collagen deposition [31] accompanied by decreased compliance and diastolic dysfunction, as previously discussed. Disruption of cardiac lymphatics in rabbits leads to significant decreases in left ventricular ejection fraction within the first 3 months following the lymphatic obstruction. This functional loss is accompanied by sustained elevations in levels of circulating plasma endothelin-1 and angiotensin II [45].

Development of pulmonary hypertension is an inevitable consequence of both acute and chronic left ventricular dysfunction and can be a prominent sequela of heart failure [8]. With increased resistance in the right ventricular outflow tract, the central venous pressure rises, reducing myocardial lymph transit into the central venous system. Increased lymphatic pressure is ultimately conveyed to the myocardial lymphatics [8]. Coronary sinus pressure is also affected [8], increasing coronary microvascular pressure, interstitial fluid filtration into the interstitium, and myocardial edema. Secondary right heart failure exacerbates the perturbations [8]. Conversely, increased pulmonary blood flow, as occurs in some forms of congenital heart defects, leads to functional and structural aberrations in lung lymphatics [46].

The impact of these various mechanisms is dependent upon the existing demands on the cardiac lymphatic vasculature [8]. Experimental elevation of coronary sinus pressure in chronic disease models produces measurable increases in myocardial water content [47] significantly earlier than a comparable intervention in healthy animal subjects [8]. Thus, the burden of additional edematous forces is more apparent when auto-regulatory mechanisms are already taxed. Coronary vascular resistance is elevated in a direct linear relationship with myocardial edema [48] and can be conceptualized as a compensatory mechanism. Therefore, the contribution of the cardiac lymphatics to the propagation of chronic myocardial edema must not be overlooked. Loss of compensatory mechanisms is likely to play an important role in the evolution of congestive heart failure, independent of the primary pathogenesis. More recent work has shown that the heart responds by increasing myocardial lymphangiogenesis from the existing vascular tree, as opposed to de novo growth from circulating progenitors [49]. Moreover, the patterns of microvascular remodeling occurring during dilated cardiomyopathy differ from those of ischemic cardiomyopathy [50].

These phenomena have been studied in human tissues derived from patients with terminal heart failure due to ischemic (ICM) and dilated (DCM) cardiomyopathy [50]. When compared to control donor heart tissues, DCM hearts demonstrate a significantly higher density of LYVE-1 positive lymphatics ($p < 0.05$), whereas no difference was seen for other markers. ICM hearts display a significantly higher density of D2-40 positive lymphatics ($p < 0.01$) and a lower density of VEGFR-2 capillaries compared to control ($p < 0.05$). Further research may help to elucidate the impact of extracellular matrix composition and VEGF-related angiogenesis on the myocardial microvasculature at various stages of heart failure.

Cardiac Transplantation

As a therapeutic intervention, cardiac transplantation poses multiple challenges to the maintenance of lymphatic function within the heart. Surgical disruption of the cardiac lymphatic vasculature during transplantation likely contributes to allograft failure through various mechanisms already discussed in this chapter, including vasculopathy and myocardial edema [51]. As is the case for the evolution of myocardial infarction, hypoxia reduces cardiac output and causes myocardial edema. Commensurate with the attempts of the autonomic nervous system to conserve perfusion capacity, there is a concurrent increase in the central venous resistance. Through similar mechanisms, experimentally induced increases in coronary sinus pressure also promote formation of myocardial edema. Cardiopulmonary bypass and cardioplegic arrest further promote myocardial edema by decreasing plasma colloid osmotic pressure and increasing plasma filtration while lymph flow diminishes [4]. The manipulations during organ procurement and transportation contribute only slightly to the overall degree of edema observed. Significant intracardiac interstitial fluid accumulation is seen only after reperfusion, reflecting the suppression of Starling equilibrium variables during cardioplegic arrest [52]. Echocardiographic studies suggest that the additional interstitial fluid distends the left ventricular wall, with spontaneous resolution over 3 months [53]. Persistence (or re-accumulation) of myocardial edema fluid precedes the cellular responses of acute rejection [54]; considered in this light, edema detection could be considered as a prognostic surrogate for post-transplant patients. Impaired lymphatic flow across the myocardium further predisposes the pharmacologically immunosuppressed system to infection, whereby both host and graft vessels become damaged by the pathological responses. In particular, cardiac allograft vasculopathy may be a long-term consequence of lymphatic stasis [55]. Moreover, in the absence of transplantation, intramural coronary arteries are remarkably spared from atherosclerosis; it is only in the context of cardiac transplantation that significant intramural coronary atherosclerosis is encountered. Disruption of the transmural plexus of lymphatics surrounding the intramural coronary arteries may explain this phenomenon [4].

The utility of hyaluronidase to limit myocardial edema has been demonstrated in an experimental model of acute rejection following heart transplantation [56]. The underlying mechanisms are not specific to transplantation, but likely apply to the more general phenomenon of myocardial edema. Analogous to observations in MI [28], administration of hyaluronidase during cardioplegic arrest promotes active drainage of cardiac lymph and reduces interstitial edema. With decreased myocardial water content, endpoint surrogates of aerobic metabolism and post-ischemic recovery of cardiac function improve [37].

Recent histological evidence corroborates the physiologic studies of the lymphatic vasculature in cardiac transplantation. This work is aided by the discovery and use of specific immunohistochemical markers of lymphatic endothelial cells [3]. Two of the best recognized markers, LYVE-1 and Prox1, are down-regulated following heart transplantation [57], while expression of VEGFR-3, the cognate receptor for the pro-lymphangiogenic factors VEGF-C and VEGF-D, remains unaltered. LYVE-1 is a transmembrane glycoprotein receptor for the extracellular matrix glycosaminoglycan, hyaluronan, among other molecules including osteopontin, collagens, and matrix metalloproteinases. Functionally, these molecules play a role in a variety of cellular processes, including lymphocyte migration and activation, hematopoiesis, and tumor metastasis [58, 59]. Although LYVE-1 is closely associated with lymphatic endothelium early in development and throughout maturity, the precise function of the receptor remains unknown (beyond its putative role in hyaluronan homeostasis) [60]. The primary receptor for hyaluronic acid, CD44, is known to facilitate cell migration by removing pericellular matrix surrounding fibroblast and epithelial cells, suppressing intercellular adhesion during wound healing, inflammation, and tumor progression [61]. Thus, LYVE-1 may play a functional role in both physiological and pathological lymphangiogenesis through its ability to transport hyaluronic acid across the lymphatic vessel wall. Nearly exclusive localization to the lymphatic endothelium throughout the vasculature, together with convenient assay techniques, renders LYVE-1 aptly as useful a molecular and histochemical marker of lymphatics, helping to distinguish them from blood vasculature. In addition, prospero-related homeobox 1 (Prox1), a nuclear transcription factor, is exclusively expressed on cells of committed lymphatic lineage during development [62]. This is in contrast to LYVE-1 and VEGFR3, which are also expressed on a limited population of non-lymphatic endothelial cells [63]. Although Prox1 is necessary and sufficient for lymphatic commitment [63], the molecular milieu in which Prox1 operates is not known; both downstream initiating and regulatory factors and other upstream requisites or supplemental events are still under investigation [64, 65]. In the context of cardiac transplantation, the postsurgical decrease in the density of LYVE-1 and Prox1, with preserved levels of VEGFR-3, suggests that the phenotype of the lymphatics within the graft is altered from wild type [57]. Alternatively, it is conceivable that a reduction in the population of lymphatic endothelial cells induces a compensatory up-regulation of the VEGFR-3 expression and, thus, the lymphangiogenic signal. It is perhaps of greater significance that VEGFR-3-positive cell density inversely correlates with the observed incidence of graft incidence [57]; observations within an experimental animal model suggest that the resumption of adequate immune modulation leads to rapid restoration of inner lymphatic vessels [66]. Further investigation of the various converging biological processes (myocardial fluid regulation, lymphocyte trafficking, and inflammation) warrants further investigation. Such studies are likely to lead to enhanced mechanistic insights and therapeutic approaches.

Future Perspectives

Molecular and ultrastructural study of the lymphatic vasculature is still in its infancy. From the foregoing discussion, it should be evident that, in future, individuals with a variety of cardiovascular pathologies may benefit directly from the enhanced insights to be gained from research into the lymphatic mechanisms that contribute to the genesis and propagation of these and other systemic diseases [67]. Progress will entail enhanced imaging modalities for the dynamic function of lymphatic vasculature within the cardiovascular structures, perhaps aided by the application of molecular imaging using a nanotechnology approach. Exploration of the direct role of lymphatic mechanisms of lipoprotein homeostasis within the arterial wall is quite desirable and may lead to new therapeutic applications. The recent identification of lymphatic mechanisms that contribute to chronic transplant rejection in other organ systems [68–70] may have direct applicability to the treatment and prevention of cardiac allograft rejection.

References

1. Oliver G, Alitalo K (2005) The lymphatic vasculature: recent progress and paradigms. Annu Rev Cell Dev Biol 21:457–483
2. Shin WS, Szuba A, Rockson SG (2003) Animal models for the study of lymphatic insufficiency. Lymphat Res Biol 1(2):159–169
3. Nakamura K, Rockson SG (2007) Biomarkers of lymphatic function and disease: state of the art and future directions. Mol Diagn Ther 11(4):227–238

4. Mehlhorn U, Geissler HJ, Laine GA, Allen SJ (2001) Myocardial fluid balance. Eur J Cardiothorac Surg 20(6):1220–1230
5. Santos AC, de Lima JJ, Botelho MF, Pacheco MF, Sousa P, Bernardo J et al (1998) Cardiac lymphatic dynamics after ischemia and reperfusion–experimental model. Nucl Med Biol 25(7):685–688
6. Yoffey JM, Courtice FC (1970) Lymphatics, lymph and the lymphomyeloid complex. Academic Pres, New York
7. Mehlhorn U, Davis KL, Laine GA, Geissler HJ, Allen SJ (1996) Myocardial fluid balance in acute hypertension. Microcirculation 3(4):371–378
8. Laine GA, Allen SJ (1991) Left ventricular myocardial edema. Lymph flow, interstitial fibrosis, and cardiac function. Circ Res 68(6):1713–1721
9. Stewart RH, Geissler HJ, Allen SJ, Laine GA (2000) Protein washdown as a defense mechanism against myocardial edema. Am J Physiol Heart Circ Physiol 279(4):H1864–H1868
10. Miller AJ, Bruna J, Beninson J (1999) A universally applicable clinical classification of lymphedema. Angiology 50(3):189–192
11. Sun SC, Lie JT (1977) Cardiac lymphatic obstruction: ultrastructure of acute-phase myocardial injury in dogs. Mayo Clin Proc 52(12):785–792
12. Tabibiazar R, Cheung L, Han J, Swanson J, Beilhack A, An A et al (2006) Inflammatory manifestations of experimental lymphatic insufficiency. PLoS Med 3(7):e254
13. Kholova I, Dragneva G, Cermakova P, Laidinen S, Kaskenpaa N, Hazes T et al (2011) Lymphatic vasculature is increased in heart valves, ischaemic and inflamed hearts and in cholesterol-rich and calcified atherosclerotic lesions. Eur J Clin Invest 41(5):487–497
14. Predescu D, Predescu S, McQuistan T, Palade GE (1998) Transcytosis of alpha1-acidic glycoprotein in the continuous microvascular endothelium. Proc Natl Acad Sci USA 95(11):6175–6180
15. Cooke CJ, Nanjee MN, Stepanova IP, Olszewski WL, Miller NE (2004) Variations in lipid and apolipoprotein concentrations in human leg lymph: effects of posture and physical exercise. Atherosclerosis 173(1):39–45
16. Schmeisser A, Christoph M, Augstein A, Marquetant R, Kasper M, Braun-Dullaeus RC et al (2006) Apoptosis of human macrophages by Flt-4 signaling: implications for atherosclerotic plaque pathology. Cardiovasc Res 71(4):774–784
17. Enholm B, Karpanen T, Jeltsch M, Kubo H, Stenback F, Prevo R et al (2001) Adenoviral expression of vascular endothelial growth factor-C induces lymphangiogenesis in the skin. Circ Res 88(6):623–629
18. Szuba A, Skobe M, Karkkainen MJ, Shin WS, Beynet DP, Rockson NB et al (2002) Therapeutic lymphangiogenesis with human recombinant VEGF-C. FASEB J 16:U114–U130
19. Geng YJ, Libby P (1995) Evidence for apoptosis in advanced human atheroma. Colocalization with interleukin-1 beta-converting enzyme. Am J Pathol 147(2):251–266
20. Davies MJ (1990) A macro and micro view of coronary vascular insult in ischemic heart disease. Circulation 82(3 Suppl):II38–II46
21. Nakano T, Nakashima Y, Yonemitsu Y, Sumiyoshi S, Chen YX, Akishima Y et al (2005) Angiogenesis and lymphangiogenesis and expression of lymphangiogenic factors in the atherosclerotic intima of human coronary arteries. Hum Pathol 36(4):330–340
22. Van Lenten BJ, Wagner AC, Nayak DP, Hama S, Navab M, Fogelman AM (2001) High-density lipoprotein loses its anti-inflammatory properties during acute influenza a infection. Circulation 103(18):2283–2288
23. Lentsch AB, Ward PA (2000) Regulation of inflammatory vascular damage. J Pathol 190(3):343–348
24. Ross R (1999) Atherosclerosis is an inflammatory disease. Am Heart J 138(5 Pt 2):S419–S420
25. Eliska O, Eliskova M, Miller AJ (2006) The absence of lymphatics in normal and atherosclerotic coronary arteries in man: a morphologic study. Lymphology 39(2):76–83

26. Rauramaa R, Halonen P, Vaisanen SB, Lakka TA, Schmidt-Trucksass A, Berg A et al (2004) Effects of aerobic physical exercise on inflammation and atherosclerosis in men: the DNASCO Study: a six-year randomized, controlled trial. Ann Intern Med 140(12):1007–1014
27. Gloviczki P, Solti F, Szlavy L, Jellinek H (1983) Ultrastructural and electrophysiologic changes of experimental acute cardiac lymphostasis. Lymphology 16(3):185–192
28. Taira A, Uehara K, Fukuda S, Takenada K, Koga M (1990) Active drainage of cardiac lymph in relation to reduction in size of myocardial infarction: an experimental study. Angiology 41(12):1029–1036
29. Dellsperger KC, Clothier JL, Hartnett JA, Haun LM, Marcus ML (1988) Acceleration of the wavefront of myocardial necrosis by chronic hypertension and left ventricular hypertrophy in dogs. Circ Res 63(1):87–96
30. Solti F, Lengyel E, Jellinek H, Schneider F, Juhasz-Nagy A, Kekesi V (1994) Coronary arteriopathy after lymphatic blockade: an experimental study in dogs. Lymphology 27(4):173–180
31. Stolyarov VV, Lushnikova EL, Zuevskii VP, Usynin AF (2002) Morphometric analysis of the lymph system in rat heart during myocardial infarction. Bull Exp Biol Med 134(2):203–205
32. Ishikawa Y, Akishima-Fukasawa Y, Ito K, Akasaka Y, Tanaka M, Shimokawa R et al (2007) Lymphangiogenesis in myocardial remodelling after infarction. Histopathology 51(3):345–353
33. Rockson SG (2001) Lymphedema. Am J Med 110(4):288–295
34. Kong D, Kong X, Wang L (2006) Effect of cardiac lymph flow obstruction on cardiac collagen synthesis and interstitial fibrosis. Physiol Res 55(3):253–258
35. Maclean D, Fishbein MC, Maroko PR, Braunwald E (1976) Hyaluronidase-induced reductions in myocardial infarct size. Science 194(4261):199–200
36. Maroko PR, Hillis LD, Muller JE, Tavazzi L, Heyndrickx GR, Ray M et al (1977) Favorable effects of hyaluronidase on electrocardiographic evidence of necrosis in patients with acute myocardial infarction. N Engl J Med 296(16):898–903
37. Yotsumoto G, Moriyama Y, Yamaoka A, Taira A (1998) Experimental study of cardiac lymph dynamics and edema formation in ischemia/reperfusion injury–with reference to the effect of hyaluronidase. Angiology 49(4):299–305
38. Cairns JA, Holder DA, Tanser P, Missirlis E (1982) Intravenous hyaluronidase therapy for myocardial infarction in man: double-blind trial to assess infarct size limitation. Circulation 65(4):764–771
39. MILIS Study Group (1986) Hyaluronidase therapy for acute myocardial infarction: results of a randomized, blinded, multicenter trial. Am J Cardiol 57(15):1236–1243
40. Roberts R, Braunwald E, Muller JE, Croft C, Gold HK, Hartwell TD et al (1988) Effect of hyaluronidase on mortality and morbidity in patients with early peaking of plasma creatine kinase MB and non-transmural ischaemia. Multicentre investigation for the limitation of infarct size (MILIS). Br Heart J 60(4):290–298
41. Saltissi S, Robinson PS, Coltart DJ, Webb-Peploe MM, Croft DN (1982) Effects of early administration of a highly purified hyaluronidase preparation (GL enzyme) on myocardial infarct size. Lancet 1(8277):867–871
42. Repa I, Garnic JD, Hollenberg NK (1990) Myocardial infarction treated with two lymphagogues, calcium dobesilate (CLS 2210) and hyaluronidase: a coded, placebo-controlled animal study. J Cardiovasc Pharmacol 16(2):286–291
43. Nakamura K, Rockson SG (2008) The role of the lymphatic circulation in the natural history and expression of cardiovascular disease. Int J Cardiol 129(3):309–317
44. Allen SJ, Geissler HJ, Davis KL, Gogola GR, Warters RD, de Vivie ER et al (1997) Augmenting cardiac contractility hastens myocardial edema resolution after cardiopulmonary bypass and cardioplegic arrest. Anesth Analg 85(5):987–992
45. Wang YL, Wang XH, Liu YL, Kong XQ, Wang LX (2009) Cardiac lymphatic obstruction impairs left ventricular function and increases plasma endothelin-1 and angiotensin II in rabbits. Lymphology 42(4):182–187

46. Datar SA, Johnson EG, Oishi PE, Johengen M, Tang E, Aramburo A et al (2012) Altered lymphatics in an ovine model of congenital heart disease with increased pulmonary blood flow. Am J Physiol Lung Cell Mol Physiol 302(6):L530–L540
47. Pratt JW, Schertel ER, Schaefer SL, Esham KE, McClure DE, Heck CF et al (1996) Acute transient coronary sinus hypertension impairs left ventricular function and induces myocardial edema. Am J Physiol 271(3 Pt 2):H834–H841
48. Rubboli A, Sobotka PA, Euler DE (1994) Effect of acute edema on left ventricular function and coronary vascular resistance in the isolated rat heart. Am J Physiol 267(3 Pt 2):H1054–H1061
49. Dashkevich A, Bloch W, Antonyan A, Fries JU, Geissler HJ (2009) Morphological and quantitative changes of the initial myocardial lymphatics in terminal heart failure. Lymphat Res Biol 7(1):21–27
50. Dashkevich A, Bloch W, Antonyan A, Goebel H, Fries JU, Schlensak C et al (2010) Immunohistochemical study of remodeling of myocardial lymphatic and blood microvascular structures in terminal heart failure: differences between ischemic and dilated cardiomyopathy. Lymphology 43(3):110–117
51. Kong XQ, Wang LX, Kong DG (2007) Cardiac lymphatic interruption is a major cause for allograft failure after cardiac transplantation. Lymphat Res Biol 5(1):45–47
52. Slater JP, Amirhamzeh MM, Yano OJ, Shah AS, Starr JP, Kaplon RJ et al (1995) Discriminating between preservation and reperfusion injury in human cardiac allografts using heart weight and left ventricular mass. Circulation 92(9 Suppl):II223–II227
53. Hosenpud JD, Norman DJ, Cobanoglu MA, Floten HS, Conner RM, Starr A (1987) Serial echocardiographic findings early after heart transplantation: evidence for reversible right ventricular dysfunction and myocardial edema. J Heart Transplant 6(6):343–347
54. Baddoura FK, Nasr IW, Wrobel B, Li Q, Ruddle NH, Lakkis FG (2005) Lymphoid neogenesis in murine cardiac allografts undergoing chronic rejection. Am J Transplant 5(3):510–516
55. Weis M, von Scheidt W (1997) Cardiac allograft vasculopathy: a review. Circulation 96(6):2069–2077
56. Johnsson C, Hallgren R, Elvin A, Gerdin B, Tufveson G (1999) Hyaluronidase ameliorates rejection-induced edema. Transpl Int 12(4):235–243
57. Geissler HJ, Dashkevich A, Fischer UM, Fries JW, Kuhn-Regnier F, Addicks K et al (2006) First year changes of myocardial lymphatic endothelial markers in heart transplant recipients. Eur J Cardiothorac Surg 29(5):767–771
58. Banerji S, Ni J, Wang SX, Clasper S, Su J, Tammi R et al (1999) LYVE-1, a new homologue of the CD44 glycoprotein, is a lymph-specific receptor for hyaluronan. J Cell Biol 144(4):789–801
59. Prevo R, Banerji S, Ferguson DJ, Clasper S, Jackson DG (2001) Mouse LYVE-1 is an endocytic receptor for hyaluronan in lymphatic endothelium. J Biol Chem 276(22):19420–19430
60. Jackson DG (2004) Biology of the lymphatic marker LYVE-1 and applications in research into lymphatic trafficking and lymphangiogenesis. APMIS 112(7–8):526–538
61. Ji RC (2006) Lymphatic endothelial cells, lymphangiogenesis, and extracellular matrix. Lymphat Res Biol 4(2):83–100
62. Wigle JT, Oliver G (1999) Prox1 function is required for the development of the murine lymphatic system. Cell 98(6):769–778
63. Wigle JT, Harvey N, Detmar M, Lagutina I, Grosveld G, Gunn MD et al (2002) An essential role for Prox1 in the induction of the lymphatic endothelial cell phenotype. EMBO J 21(7):1505–1513
64. Francois M, Caprini A, Hosking B, Orsenigo F, Wilhelm D, Browne C et al (2008) Sox18 induces development of the lymphatic vasculature in mice. Nature 456(7222):643–647
65. Francois M, Harvey NL, Hogan BM (2011) The transcriptional control of lymphatic vascular development. Physiology (Bethesda) 26(3):146–155
66. Soong TR, Pathak AP, Asano H, Fox-Talbot K, Baldwin WM 3rd (2010) Lymphatic injury and regeneration in cardiac allografts. Transplantation 89(5):500–508

67. Rockson SG (2006) The lymphatic continuum continues. Lymphat Res Biol Spring;4(1):1–2
68. Bock F, Onderka J, Dietrich T, Bachmann B, Kruse FE, Paschke M et al (2007) Bevacizumab as a potent inhibitor of inflammatory corneal angiogenesis and lymphangiogenesis. Invest Ophthalmol Vis Sci 48(6):2545–2552
69. Kerjaschki D, Huttary N, Raab I, Regele H, Bojarski-Nagy K, Bartel G et al (2006) Lymphatic endothelial progenitor cells contribute to de novo lymphangiogenesis in human renal transplants. Nat Med 12(2):230–234
70. Kerjaschki D, Regele HM, Moosberger I, Nagy-Bojarski K, Watschinger B, Soleiman A et al (2004) Lymphatic neoangiogenesis in human kidney transplants is associated with immunologically active lymphocytic infiltrates. J Am Soc Nephrol 15(3):603–612
71. Tammela T, Saaristo A, Lohela M, Morisada T, Tornberg J, Norrmen C et al (2005) Angiopoietin-1 promotes lymphatic sprouting and hyperplasia. Blood 105(12):4642–4648
72. Gale N, Thurston G, Hackett S, Renard R, Wang Q, McClain J et al (2002) Angiopoietin-2 is required for postnatal angiogenesis and lymphatic patterning, and only the latter role is rescued by angiopoietin-1. Dev Cell 3:411–423
73. Hirakawa S, Hong YK, Harvey N, Schacht V, Matsuda K, Libermann T et al (2003) Identification of vascular lineage-specific genes by transcriptional profiling of isolated blood vascular and lymphatic endothelial cells. Am J Pathol 162(2):575–586
74. Kriehuber E, Breiteneder-Geleff S, Groeger M, Soleiman A, Schoppmann SF, Stingl G et al (2001) Isolation and characterization of dermal lymphatic and blood endothelial cells reveal stable and functionally specialized cell lineages. J Exp Med 194(6):797–808
75. Gunn MD, Tangemann K, Tam C, Cyster JG, Rosen SD, Williams LT (1998) A chemokine expressed in lymphoid high endothelial venules promotes the adhesion and chemotaxis of naive T lymphocytes. Proc Natl Acad Sci USA 95(1):258–263
76. Ebata N, Nodasaka Y, Sawa Y, Yamaoka Y, Makino S, Totsuka Y et al (2001) Desmoplakin as a specific marker of lymphatic vessels. Microvasc Res 61(1):40–48
77. Kriederman BM, Myloyde TL, Witte MH, Dagenais SL, Witte CL, Rennels M et al (2003) FOXC2 haploinsufficient mice are a model for human autosomal dominant lymphedema-distichiasis syndrome. Hum Mol Genet 12(10):1179–1185
78. Petrova TV, Karpanen T, Norrmen C, Mellor R, Tamakoshi T, Finegold D et al (2004) Defective valves and abnormal mural cell recruitment underlie lymphatic vascular failure in lymphedema distichiasis. Nat Med 10(9):974–981
79. Kajiya K, Hirakawa S, Ma B, Drinnenberg I, Detmar M (2005) Hepatocyte growth factor promotes lymphatic vessel formation and function. EMBO J 24(16):2885–2895
80. Huang XZ, Wu JF, Ferrando R, Lee JH, Wang YL, Farese RV Jr et al (2000) Fatal bilateral chylothorax in mice lacking the integrin alpha9beta1. Mol Cell Biol 20(14):5208–5215
81. Petrova T, Makinen T, Makela T, Saarela J, Virtanen I, Ferrell R et al (2002) Lymphatic endothelial reprogramming of vascular endothelial cells by the Prox-1 homeobox transcription factor. EMBO J 21:4593–4599
82. Irjala H, Johansson EL, Grenman R, Alanen K, Salmi M, Jalkanen S (2001) Mannose receptor is a novel ligand for l-selectin and mediates lymphocyte binding to lymphatic endothelium. J Exp Med 194(8):1033–1042
83. Yuan L, Moyon D, Pardanaud L, Breant C, Karkkainen M, Alitalo K et al (2002) Abnormal lymphatic vessel development in neuropilin 2 mutant mice. Development 129:4797–4806
84. Ayadi A, Zheng H, Sobieszczuk P, Buchwalter G, Moerman P, Alitalo K et al (2001) Net-targeted mutant mice develop a vascular phenotype and up-regulate egr-1. EMBO J 20(18):5139–5152
85. Harvey NL, Srinivasan RS, Dillard ME, Johnson NC, Witte MH, Boyd K et al (2005) Lymphatic vascular defects promoted by Prox1 haploinsufficiency cause adult-onset obesity. Nat Genet 37(10):1072–1081
86. Breiteneder-Geleff S, Soleiman A, Kowalski H, Horvat R, Amann G, Kriehuber E et al (1999) Angiosarcomas express mixed endothelial phenotypes of blood and lymphatic capillaries: podoplanin as a specific marker for lymphatic endothelium. Am J Pathol 154(2):385–394

87. Schacht V, Ramirez MI, Hong YK, Hirakawa S, Feng D, Harvey N et al (2003) T1alpha/podoplanin deficiency disrupts normal lymphatic vasculature formation and causes lymphedema. EMBO J 22(14):3546–3556
88. Pennisi D, Gardner J, Chambers D, Hosking B, Peters J, Muscat G et al (2000) Mutations in Sox18 underlie cardiovascular and hair follicle defects in ragged mice. Nat Genet 24(4):434–437
89. Abtahian F, Guerriero A, Sebzda E, Lu MM, Zhou R, Mocsai A et al (2003) Regulation of blood and lymphatic vascular separation by signaling proteins SLP-76 and Syk. Science 299(5604):247–251
90. Jeltsch M, Kaipainen A, Joukov V, Meng X, Lakso M, Rauvala H et al (1997) Hyperplasia of lymphatic vessels in VEGF-C transgenic mice. Science 276(5317):1423–1425
91. Karkkainen MJ, Ferrell RE, Lawrence EC, Kimak MA, Levinson KL, McTigue MA et al (2000) Missense mutations interfere with VEGFR-3 signalling in primary lymphoedema. Nat Genet 25(2):153–159
92. Dumont DJ, Jussila L, Taipale J, Lymboussaki A, Mustonen T, Pajusola K et al (1998) Cardiovascular failure in mouse embryos deficient in VEGF receptor-3. Science 282(5390):946–949
93. Kaipainen A, Korhonen J, Mustonen T, van Hinsbergh VW, Fang GH, Dumont D et al (1995) Expression of the fms-like tyrosine kinase 4 gene becomes restricted to lymphatic endothelium during development. Proc Natl Acad Sci USA 92(8):3566–3570
94. Karkkainen MJ, Saaristo A, Jussila L, Karila KA, Lawrence EC, Pajusola K et al (2001) A model for gene therapy of human hereditary lymphedema. Proc Natl Acad Sci USA 98(22):12677–12682

Chapter 5
The Lymphatics in Normal and Pathological Heart Valves

Ivana Kholová, Galina Dragneva, and Seppo Ylä-Herttuala

Abstract Despite the importance of the myocardial lymphatic vasculature in many pathological conditions, little information is readily available about heart valvular lymphatic vessels in normal and pathological conditions. Before the onset of specific antibodies, researchers mainly performed dye and hydrogen peroxide injection studies in animal hearts in order to visualize the cardiac lymphatics. In the era of specific antibodies, lymphatic vessels were described in normal human valves. In pathological valves, the highest number of lymphatics was found in valves with infective endocarditis where they accounted for nearly 100 % of all vessels in extracellular matrix-rich areas whereas inflammatory cell-rich areas were more prone to angiogenesis. An increased number of lymphatics was also found in cases of degenerative calcified stenosis and myxoid degeneration whereas the number was unchanged in fibrotic valves. Knowledge of the lymphatics in cardiac valves and their changes under various pathological conditions is also important for the further development of various treatment strategies. Certain drugs or gene therapy techniques could potentially influence lymphatic vessel densities in these regions.

Keywords Lymphatic vessel • Heart valve • Endocarditis • Valvular disease • Lymphatic marker • Lymphangiogenesis • Human heart

---

I. Kholová, MD, PhD
Fimlab Laboratories, Pathology, Tampere University Hospital,
Biokatu 4, 33520 Tampere, Finland
e-mail: ivana.kholova@sll.fimnet.fi

G. Dragneva, MSc • S. Ylä-Herttuala, MD, PhD (✉)
Department of Biotechnology and Molecular Medicine, A.I. Virtanen Institute,
University of Eastern Finland, Neulaniementie 2, 70211 Kuopio, Finland
e-mail: galina.dragneva@uef.fi; Seppo.Ylaherttuala@uef.fi

G. Karunamuni (ed.), *The Cardiac Lymphatic System: An Overview*,
DOI 10.1007/978-1-4614-6774-8_5, © Springer Science+Business Media New York 2013

Historical Perspective

There exists a large body of evidence that supports the presence of a lymphatic network in the cardiac valves. Over 150 years ago, Eberth and Belajeff [1] described a subendocardial plexus of lymphatic vessels extending into the atrioventricular and semilunar valves both in humans and other mammals. In 1924, Aagaard showed lymphatics entering the leaflets of atrioventricular valves in animals, but was unable to demonstrate lymphatics in the cardiac valves of man [2]. Later, other researchers such as Patek [3] and Bradham et al. [4] failed to demonstrate lymphatics by using injection techniques. In the beginning of the 1960s, Miller et al. visualized lymphatics in normal and diseased canine mitral valves [5]. Noguchi et al. used a series of techniques to reveal the morphology of atrioventricular valve lymphatics in dogs [6]. Some of these techniques will be described in the following sections.

Animal Studies

The overwhelming majority of animal experiments have been performed in canine hearts. In 1961, pioneering works of Miller et al. showed thin-walled channels consisting of an endothelial lining and sparse surrounding connective tissue in the anterior mitral leaflet of dogs. Miller et al. proved that these channels were more numerous and larger in caliber after chronic impairment of cardiac lymph flow. Furthermore, they speculated that the channels were formed either by opening preexisting collaterals or by the growth of new channels into the valve leaflets [5].

The impact of cardiac lymph flow impairment on atrioventricular valves was also demonstrated by Symbas et al. The atrioventricular valves, particularly the tricuspid valves, revealed thickening caused by the accumulation of amorphous myxoid material composed mainly of hyaluronic acid and chondroitin sulfates. Those changes led to a significant change in the valve architecture represented by fibrosis [7]. Ullal et al. showed that blocking cardiac lymph flow led to the dilatation of lymphatics, myxoid deposition, and mild fibrosis in valves [8].

Using the hydrogen peroxide technique and stereomicroscopy, Johnson and Blake demonstrated the presence of lymphatics in mitral and tricuspid valves of normal dogs and pigs but not in aortic or pulmonary valves [9].

Noguchi applied four techniques, including India ink injection, a hydrogen peroxide technique, light microscopy, and electron microscopy. Toluidine blue-stained Epon sections revealed lymphatics as thin-walled vessels with irregular contours of 70–150 μm diameter. Lymphatic capillaries in adult dogs and puppies were noted in the subendocardial layer and not in the connective tissue plate of the atrioventricular valves. Lymphatic vessel number varied among the different cusps, with lymphatics being most numerous in the anterior cusp of the mitral valve, only sporadically present in the septal cusp of the tricuspid valve, and relatively sparse in the other cusps of the atrioventricular valves [6].

Methodological Approaches for Studying Lymphatics in Valves

Until the generation of specific antibodies for lymphatic vessels, knowledge of mammalian cardiac lymphatics was extrapolated mainly from animal studies conducted with various injection techniques. However, injection techniques do not allow for analysis of deeper vasculature [10].

For decades, dye injection has been a standard technique. Dye, e.g., India ink or Evans Blue, is injected with the aid of glass capillaries through the opened pericardium. After cardiac collecting lymphatic vessels have been identified, they are catheterized and dye is reinjected to visualize terminal lymphatic pathways. After fixation, the structures can be observed grossly or with the use of a stereomicroscope or dissecting microscope [6, 11, 12].

Alternatively, the hydrogen peroxide technique can be applied. Hydrogen peroxide initiates an oxidoreduction reaction with catalase and peroxidase in the lymph, thus producing oxygen and water. The released oxygen causes distention of lymphatics. One percent solution of hydrogen peroxide is applied either topically with a cotton-tipped applicator or a whole heart can be immersed in the solution. The reaction can be enhanced either through pre-refrigeration or prefixation [9, 13].

Electron microscopy enables the detailed study of the fine structural architecture of lymphatics in human hearts. Using this method, morphological information can be retrieved from half-thin toluidine blue-stained sections [6, 12].

Currently, antibodies specific for the lymphatic vasculature allow the precise evaluation of these structures. Some commonly used antibodies are podoplanin/D2-40, LYVE-1, and Prox-1. D2-40 is a recently developed monoclonal antibody raised against a M2A antigen [14, 15]. D2-40 specifically recognizes human podoplanin. Podoplanin/D2-40 is a mucin-type transmembrane glycoprotein which was originally found on the surface of rat glomerular epithelial cells (podocytes), and loss of podoplanin has been linked to the flattening of foot processes that occurs in glomerular diseases [16]. Podoplanin is specifically expressed in the endothelium of lymphatic capillaries, but not in the blood vasculature. In normal skin and kidney, podoplanin is co-localized with VEGFR3/FLT4, yet another marker for lymphatic endothelial cells.

LYVE-1 (lymphatic vessel endothelial receptor-1) is a CD44 homolog found primarily on lymphatic endothelial cells [17]. Potential roles for LYVE-1 have been suggested in hyaluronan transport and turnover, or in promoting hyaluronan localization to the surface of the lymphatic endothelium. As a marker of lymphoid tissues and/or lymphangiogenesis, LYVE-1 is expressed on both the lumenal and ablumenal surfaces of lymphatic endothelium and also on hepatic blood sinusoidal endothelia [17, 18]. LYVE-1 also stains macrophages and adipocytes to some extent [19].

Prox1, the homolog of *Drosophila* prospero, is a homeobox-containing transcription factor that binds and functions as a co-receptor of liver receptor homolog 1 (LRH1/NR5A2). It is a specific marker for lymphatic endothelial cells expressed in nuclei [20].

Pan-endothelial marker CD31, also known as platelet endothelial cell adhesion molecule, is expressed in lymphatic endothelial cells as well [21]. CD34 is a transmembrane sialomucin protein that is expressed in hematopoietic and vascular-associated tissue [22]. CD34 is used as a blood vasculature marker; however, it was also detected in intratumoral lymphatics in colon, breast, lung, and skin tumors [23] and pleural lymphatics in lymphangiomatosis [24].

Normal Valves

Lymphatic vessels were observed in normal human heart valves in several studies [9, 21, 25]. Valvular lymphatic vessels are thin-walled with irregular lumens, their diameter being 70–150 μm, and they contain cardiac lymph intraluminally. Lymphatic endothelial cells are flat except in the area where the nucleus is located and usually overlap each other without tight junctions. The basement membrane is often discontinuous or even absent, while surrounding pericytes are sparse in number. The majority of vessels in normal cardiac valves revealed α-SMA positivity (Fig. 5.1d) [21]. Typically, a network of anchoring filaments and adjacent collagen fibers can be found in the perivascular space [6, 21, 26, 27].

Ultrastructurally, the cytoplasm of lymphatic endothelial cells contains a large proportion of small vesicles, microfilaments, endoplasmic reticuli, and lysosomes, while the Golgi apparatus is often inconspicuous. Adjacent endothelial cells are connected via end-to-end adhesion, overlapping cytoplasmic processes, and fork-like interlocking [6].

The highest density of lymphatic vessels was seen in the basal part of the valves. Scattered lymphatic vessels were also found in the peripheral part of the valves. There were no visible differences among leaflets. Lymphatic vessels were found in normal valves at an average density of 6–11 vessels/mm^2: in mitral valves (7.0 vessels/mm^2), in tricuspid valves (11.0 vessels/mm^2), and in pulmonary valves (8.0 vessels/mm^2) (Fig. 5.1a–c). Surprisingly, we were unable to detect lymphatics in the aortic valves of normal hearts probably due to sampling; however, they were observed in nonpathological aortic valves of infarcted (6.0 vessels/mm^2), fibrotic (6.0 vessels/mm^2), and hypertrophied hearts (7.0 vessels/mm^2). A similar density of lymphatics was also found in the mitral, tricuspid, and pulmonary valves of infarcted, fibrotic, and hypertrophied hearts [21].

In total, lymphatics formed up to 37 % of all vasculature in the normal cardiac valves. The lymphatic portion of all vasculature in the valves was higher in comparison to other heart compartments, i.e., ventricles and atria (Table 5.1).

Fig. 5.1 (continued) (**k**) Lymphatic vessels in myxoid degeneration in aortal valve featured sprouting and branching; however, there were only a few lymphatics in the valve. (**l**) Only a few lymphatic vessels were found in rheumatic fibrohyalinosis in a mitral valve. (**m**) The graph shows the number of lymphatic (podoplanin positive) vessels/mm^2 in normal and pathological valves in adult human hearts. Immunostaining with podoplanin (**a–c, e, g, i, k, l**), α-SMA (**d**), and CD31 (**f, h, j**). Scale bars: 200 μm (**a, c–h, j, l**), 100 μm (**b, i, k**) (Adapted from [21], with permission)

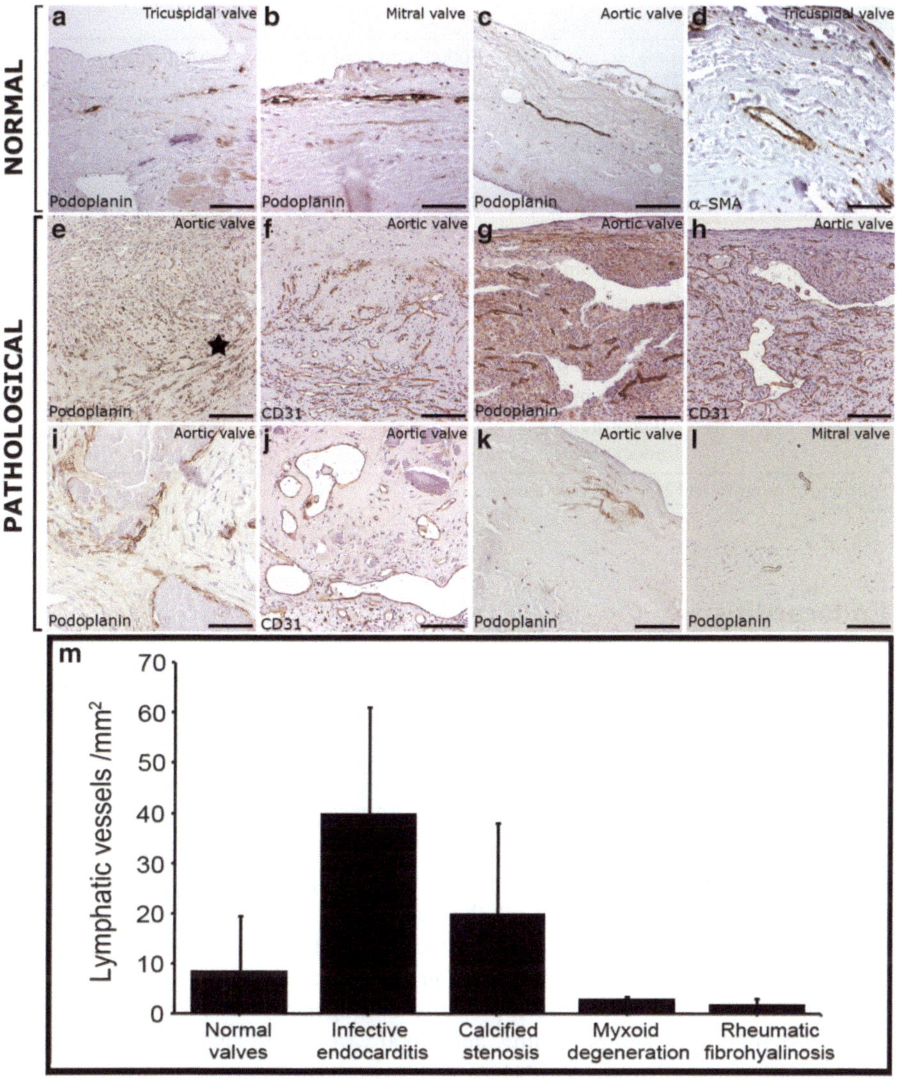

Fig. 5.1 Lymphatic and blood vessels in normal and pathological valves. (**a**) Normal tricuspidal valves revealed accumulation of lymphatic vessels in the proximal part of the valve. (**b**) However, lymphatic vessels were located also in the subendocardial part as shown, e.g., in the mitral valve. (**c**) Aortic valve from infarcted heart revealed lymphatic vessels in the middle of the leaflet. (**d**) The majority of vessels in the valves revealed α-SMA positivity (picture from tricuspidal valve). (**e–h**) Aortic valve. (**e**) Infective endocarditis contained the highest increase in the number of lymphatics. In some areas D2-40-positive lymphatics accounted for nearly 100 % of all vessels (*star*). (**f**) Corresponding area with focally dilated CD31-positive vessels. Note that areas rich in extracellular matrix contain more lymphatics (**e**) than inflammatory cell-rich areas. (**g**) Another valve leaflet with lymphatics intermingled among blood vessels in inflammatory cell-rich area. (**h**) CD31-positive vessels were predominant in inflammatory cell-rich area. (**i**) Degenerative calcified aortic valve revealed podoplanin-positive lymphatic vessels in the vicinity of calcification. (**j**) Note dilated CD31-positive blood vessels in cellular areas in degenerative calcified aortic valve.

Table 5.1 Lymphatic vessels in normal ventricles and atria

		Podoplanin-positive vessels (per mm²)	Podoplanin-positive vessels (% of all vessels)
Left ventricle	Epicardium	27.7±9.2	1.33
	Myocardium	24.7±8.3	0.53
	Endocardium	32.7±10.9	1.57
Right ventricle	Epicardium	24.0±9.6	2.20
	Myocardium	25.5±8.9	0.55
	Endocardium	18.5±7.6	1.17
Left atrium	Epicardium	22.0±7.9	1.99
	Myocardium	12.5±6.3	0.26
	Endocardium	9.5±7.8	1.09
Right atrium	Epicardium	4.0±2.9	0.55
	Myocardium	4.3±2.3	0.15
	Endocardium	0	0

Source: Adapted from [21], with permission

Endocarditis

Endocarditis can be either infective or noninfective. Cardiac and vascular abnormalities predispose to infective endocarditis, but it also develops in normal valves. Infective endocarditis is the infection of heart valves, with mural endocardium forming thrombotic debris and vegetations of microorganisms. Most cases are bacterial, but fungi, chlamydia, and rickettsiae are also rare causative agents [28]. The highest number of lymphatics was found in biopsy samples from infective endocarditis in comparison to the normal and other pathologically involved valves. With a lymphatic density of 61 vessels/mm^2, they accounted for nearly 100 % of all vessels in certain areas. There was a clear difference between areas with angiogenesis and lymphangiogenesis. Whereas lymphatics grew in areas rich in extracellular matrix, inflammatory cell-rich areas were prone to angiogenesis. The lymphatic vessels were partially dilated and branching, and blood vasculature also revealed dilatation and branching (Fig. 5.1e–h, m) [21].

Pathological changes in lymphatic vasculature in cases of endocarditis have already been speculated on by Miller et al. [5]: lymphatic obstruction predisposes to infection and inflammation. Those processes further embarrass lymph flow and enhance fibrosis. Impairment of lymphatic vasculature was recently described in various infective and inflammatory diseases, when contractile function is embarrassed. This might then decrease lymph flow and thus influence lymphatic function [29]. In organ failure, inflammation is also accompanied by dysfunction of lymphatic pumping and impairment of lymph flow [29]. Lymphatic vessels furthermore actively regulate inflammatory responses by participating in leukocyte recirculation [30]. Thus, the role of lymphatic vasculature in all inflammatory processes including endocarditis is important.

Lymphangiogenesis in endocarditis is also accompanied by angiogenesis (Fig. 5.1f, h). Recently, it has been shown that vascular formation patterns are dependent on the local microenvironment. Vascular endothelial growth factors (VEGFs) were found to be overexpressed in endocarditis valves [31]. Inflammatory cells are the main source of VEGFs in endocarditis. Geographical differences in areas with the highest lymphatic and blood vessel growth support the role of the local microenvironment in both lymphangiogenesis and angiogenesis.

Valvular Disease

As a result of valvular disease, valvular involvement can include stenosis, insufficiency (regurgitation or incompetence), or both. Major etiologies are postinflammatory scarring including rheumatic heart disease, degenerative diseases including senile involvement, and autoimmune diseases [28].

Pathological valves in valvular diseases are accompanied by lymphatic growth to a lesser extent than in infective endocarditis. Aortic calcified stenosis and myxoid degeneration revealed also either total or local increases in lymphatic vessel densities, respectively [21].

In aortic valve degenerative calcified stenosis, lymphatics were localized around calcified noncellular focuses (Fig. 5.1i, j). Focally, vessel density reached up to 40 vessels/mm^2, and the lymphatics were slightly dilated [21]. Soini et al. showed unevenly distributed angiogenesis in nonrheumatic aortic valve stenosis associated with inflammatory cells and suggested that growth factors expressed by inflammatory cells contribute to angiogenesis [32]. As mentioned above, lymphangiogenesis is often accompanied by angiogenesis, but differences in local microenvironment influence the pattern of vessel growth.

Myxoid degeneration of mitral valves was accompanied by less abundant lymphatics (Fig. 5.1k). Nevertheless, there were still more lymphatics found focally in myxoid areas compared to normal valves. Moreover, neither dilatation nor branching was observed [21].

To date, postrheumatic fibrohyalinosis has been the only documented valvular disease where there has not been an increase in lymphatic vessels. In this case, the number of lymphatics in fibrotic areas was the same as in normal valves, and the majority of vessels in these fibrotic areas were CD31-positive blood vessels (Fig. 5.1l) [21].

Conclusions

Recently, with the development of lymphatic endothelium-specific antibodies, it has been possible to study lymphatic vasculature in normal and pathologically changed human hearts. Lymphatic vessels were shown to be present both in normal and diseased valves. Striking new findings have been observed in some pathological valves, where lymphatic vessels formed the highest proportion of all vasculature.

The highest number of lymphatics in valves was found in cases of infective endocarditis where they accounted for nearly 100 % of all vessels in extracellular matrix-rich areas whereas inflammatory cell-rich areas were more prone to angiogenesis. An increased number of lymphatics were also found in degenerative calcified stenosis and myxoid degeneration whereas the number was unchanged in postrheumatic fibrotic valves [21].

The role of lymphangiogenesis in pathological valves is significant. The lymphatic vasculature is involved in the pathogenesis of infective and inflammatory diseases. Furthermore, an increase in lymphatic vessels was detected also in degenerative diseases such as calcified stenosis which is pathogenetically close to atherosclerosis.

The knowledge of the lymphatics in cardiac valves and their changes under various pathological conditions is essential for the further development of various treatment strategies, such as drugs or gene therapy techniques that could influence lymphatic vessel densities to reach desired levels.

References

1. Eberth CJ, Belajeff A (1866) Uber die Lymphagefässe des Herzens. Arch Path Anat 37:124–131
2. Aagaard OC (1924) Les Vaisseaux Lymphatique du Coeur Chez L'Homme et Chez Quelques Mammiferes. Levin and Munkegaard, Copenhagen
3. Patek P (1939) Morphology of the lymphatics of the mammalian heart. Am J Anat 64:203–234
4. Bradham RR, Parker EF, Greene WB (1973) Lymphatics of the atrioventricular valves. Arch Surg 106(2):210–213
5. Miller AJ, Pick R, Katz LN (1961) Lymphatics of the mitral valve of the dog. Demonstration and discussion of the possible significance. Circ Res 9:1005–1009
6. Noguchi T, Shimada T, Nakamura M, Uchida Y, Shirabe J (1988) The distribution and structure of the lymphatic system in dog atrioventricular valves. Arch Histol Cytol 51(4):361–370
7. Symbas PN, Schlant RC, Gravanis MB, Shepherd RL (1969) Pathologic and functional effects on the heart following interruption of the cardiac lymph drainage. J Thorac Cardiovasc Surg 57(4):577–584
8. Ullal SR, Kluge TH, Gerbode F (1972) Functional and pathologic changes in the heart following chronic cardiac lymphatic obstruction. Surgery 71(3):328–334
9. Johnson RA, Blake TM (1966) Lymphatics of the heart. Circulation 33(1):137–142
10. Eliskova M, Eliska O (1992) How lymph is drained away from the human papillary muscle: anatomical conditions. Cardiology 81(6):371–377
11. Riquet M, Le Pimpec BF, Souilamas R, Hidden G (2002) Thoracic duct tributaries from intrathoracic organs. Ann Thorac Surg 73(3):892–898; discussion 898–899
12. Sacchi G, Weber E, Agliano M, Cavina N, Comparini L (1999) Lymphatic vessels of the human heart: precollectors and collecting vessels. A morpho-structural study. J Submicrosc Cytol Pathol 31(4):515–525
13. Parke WW, Michels NA (1963) A method for demonstrating subserous lymphatics with hydrogen peroxide. Anat Rec 146:165–171

14. Bailey D, Baumal R, Law J, Sheldon K, Kannampuzha P, Stratis M et al (1986) Production of a monoclonal antibody specific for seminomas and dysgerminomas. Proc Natl Acad Sci USA 83(14):5291–5295
15. Marks A, Sutherland DR, Bailey D, Iglesias J, Law J, Lei M et al (1999) Characterization and distribution of an oncofetal antigen (M2A antigen) expressed on testicular germ cell tumours. Br J Cancer 80(3–4):569–578
16. Breiteneder-Geleff S, Matsui K, Soleiman A, Meraner P, Poczewski H, Kalt R et al (1997) Podoplanin, novel 43-kd membrane protein of glomerular epithelial cells, is down-regulated in puromycin nephrosis. Am J Pathol 151(4):1141–1152
17. Prevo R, Banerji S, Ferguson DJ, Clasper S, Jackson DG (2001) Mouse LYVE-1 is an endocytic receptor for hyaluronan in lymphatic endothelium. J Biol Chem 276(22):19420–19430
18. Mouta Carreira C, Nasser SM, di Tomaso E, Padera TP, Boucher Y, Tomarev SI et al (2001) LYVE-1 is not restricted to the lymph vessels: expression in normal liver blood sinusoids and down-regulation in human liver cancer and cirrhosis. Cancer Res 61(22):8079–8084
19. Florez-Vargas A, Vargas SO, Debelenko LV, Perez-Atayde AR, Archibald T, Kozakewich HP et al (2008) Comparative analysis of D2-40 and LYVE-1 immunostaining in lymphatic malformations. Lymphology 41(3):103–110
20. Wigle JT, Oliver G (1999) Prox1 function is required for the development of the murine lymphatic system. Cell 98(6):769–778
21. Kholova I, Dragneva G, Cermakova P, Laidinen S, Kaskenpaa N, Hazes T et al (2011) Lymphatic vasculature is increased in heart valves, ischaemic and inflamed hearts and in cholesterol-rich and calcified atherosclerotic lesions. Eur J Clin Invest 41(5):487–497
22. Nielsen JS, McNagny KM (2008) Novel functions of the CD34 family. J Cell Sci 121 (pt 22):3683–3692
23. Fiedler U, Christian S, Koidl S, Kerjaschki D, Emmett MS, Bates DO et al (2006) The sialomucin CD34 is a marker of lymphatic endothelial cells in human tumors. Am J Pathol 168(3):1045–1053
24. Eom M, Choi YD, Kim YS, Cho MY, Jung SH, Lee HY (2007) Clinico-pathological characteristics of congenital pulmonary lymphangiectasis: report of two cases. J Korean Med Sci 22(4):740–745
25. Johnson RA (1969) The lymphatic system of the heart. Lymphology 2(3):95–108
26. Alitalo K, Carmeliet P (2002) Molecular mechanisms of lymphangiogenesis in health and disease. Cancer Cell 1(3):219–227
27. Karkkainen MJ, Alitalo K (2002) Lymphatic endothelial regulation, lymphoedema, and lymph node metastasis. Semin Cell Dev Biol 13(1):9–18
28. Schoen FJ (2005) The heart. In: Kumar V, Abbas AK, Fausto N (eds) Robbins and Cotran pathologic basis of diseases. Elsevier Saunders, Philadelphia, pp 555–618
29. von der Weid PY, Muthuchamy M (2010) Regulatory mechanisms in lymphatic vessel contraction under normal and inflammatory conditions. Pathophysiology 17(4):263–276
30. Karpanen T, Alitalo K (2008) Molecular biology and pathology of lymphangiogenesis. Annu Rev Pathol 3:367–397
31. Kholova I, Laidinen S, Dragneva G, Hajkova P, Steiner I, Kosma V (2007) Lymphatic vessels, lymphatic vascular endothelial growth factors and their receptors in human heart valves. Virchows Arch 451:127
32. Soini Y, Salo T, Satta J (2003) Angiogenesis is involved in the pathogenesis of nonrheumatic aortic valve stenosis. Hum Pathol 34(8):756–763

Chapter 6
Therapeutic Applications Targeting the Cardiac Lymphatics in Heart Disease

Jae-Hyeong Park and Jin-Ok Jeong

Abstract The lymphatic system performs important roles in various essential body functions, such as fluid homeostasis, trafficking of immune cells, and intestinal lipid absorption. Despite its important roles, the lymphatic system of the heart has largely been overlooked due to its grossly invisible nature. The cardiac lymphatic system plays important roles in myocardial fluid homeostasis, controlling inflammation and infection. Because this system can contribute to various stages of myocardial infarction and inflammation, the study of this lymphatic system can elucidate the pathologic process of many cardiovascular diseases. Also, the role of the cardiac lymphatic system should be considered in the management of patients with cardiac diseases with increased risk of myocardial edema such as myopericarditis or heart failure. In this chapter, we discuss how the cardiac lymphatic system can contribute to the regulation of cardiovascular disease and provide potential therapeutic applications related to this important network of vessels.

Keywords Lymphatics • Heart • Lymphangiogenesis • Fluid balance

Anatomy and Functions of the Cardiac Lymphatic System

Anatomy

The cardiac lymphatic system, consisting of drainage vessels and lymphatic capillaries, modulates myocardial fluid homeostasis between fluid filtration into the myocardial interstitium from blood capillaries and its removal via the lymphatic system [1].

J.-H. Park, MD, PhD • J.-O. Jeong, MD, PhD (✉)
Department of Internal Medicine, Chungnam National University Hospital, Regional Cardiovascular Center, 282 Munhwa-ro, Jung-ku, 302-721 Daejeon, Korea
e-mail: jaehpark@cnu.ac.kr; jojeong@cnu.ac.kr

G. Karunamuni (ed.), *The Cardiac Lymphatic System: An Overview*,
DOI 10.1007/978-1-4614-6774-8_6,

The drainage vessels are mainly situated in subepicardial spaces following branches of the coronary artery and contain many valves [2]. These vessels are primarily located over the atria and ventricles. The lymphatic capillaries are composed of a thin layer of endothelial cells and form relatively dense networks in a fishnet arrangement. These lymphatic networks are richer in the ventricles than in the atria, being present in the subepicardial myocardial and subendocardial regions. Subendocardial lymphatics can be observed in nearly all areas of the endocardium. Beneath the endocardium of the ventricular walls and the bodies of papillary muscles, the size of lymph vessels measures between 20 and 45 μm in diameter with occasional channels up to 150 μm. The lymph channels are more nearly parallel to subjacent muscle bundles and have a bulbous appearance suggesting the presence of valves. In adult human hearts, lymphatic vessels are predominantly located in the epicardium and endocardium although there are some lymphatic vessels also in the myocardium. Lymphatics with the largest caliber are located in epicardium, endocardium, and in the connective tissue within the myocardium. The subendocardial and subepicardial systems communicate through transmyocardial channels in the atrioventricular valve annuli, papillary muscles, and at random throughout the myocardium [3].

Large cardiac lymphatic vessels run with the coronary vessels in the epicardial grooves of the heart (Fig. 6.1). The principal cardiac lymphatic vessel which drains from the left ventricular and atrial muscles passes posterior to the pulmonary trunk to connect to the right paratracheal lymph nodes [4]. The drainage of cardiac lymph from the right ventricle enters to the left anterior mediastinal chain and left paratracheal lymph nodes [5]. Lymphatics of the interventricular groove go to the right paratracheal nodes and then to the large lymphatic duct [5, 6]. There are multiple systems involving both the thoracic and the right lymphatic duct for the drainage of pericardial lymphatics [7]. Further drainage from the pericardial space goes to nodes in the right and left upper mediastinum as well as to the bilateral parasternal internal thoracic lymphatic chains [6].

Functions

The major function of the cardiac lymphatic system is to maintain homeostasis in myocardial fluid balance [8]. A lymph flow obstruction can impair the drainage of interstitial fluid and proteins in the myocardium. This impairment is usually associated with increased tissue edema and subsequent fibrosis [9]. An early animal study found that acute lymphatic obstruction caused subendocardial edema and hemorrhage within 150 min after ligation of cardiac lymphatics [9]. Ultrastructurally, lymphangiectasis, myofibrillar degeneration, disruption of Z-band, and intercalated discs can be observed after acute lymphatic obstruction. Also, acute cardiac lymphatic obstruction depresses myocardial systolic and diastolic functions [10]. Chronic lymphatic obstruction can cause myocardial fibrosis and decrease myocardial contractility, and this effect has been found to be important in the pathogenesis of myocardial and valvular heart diseases [9, 11].

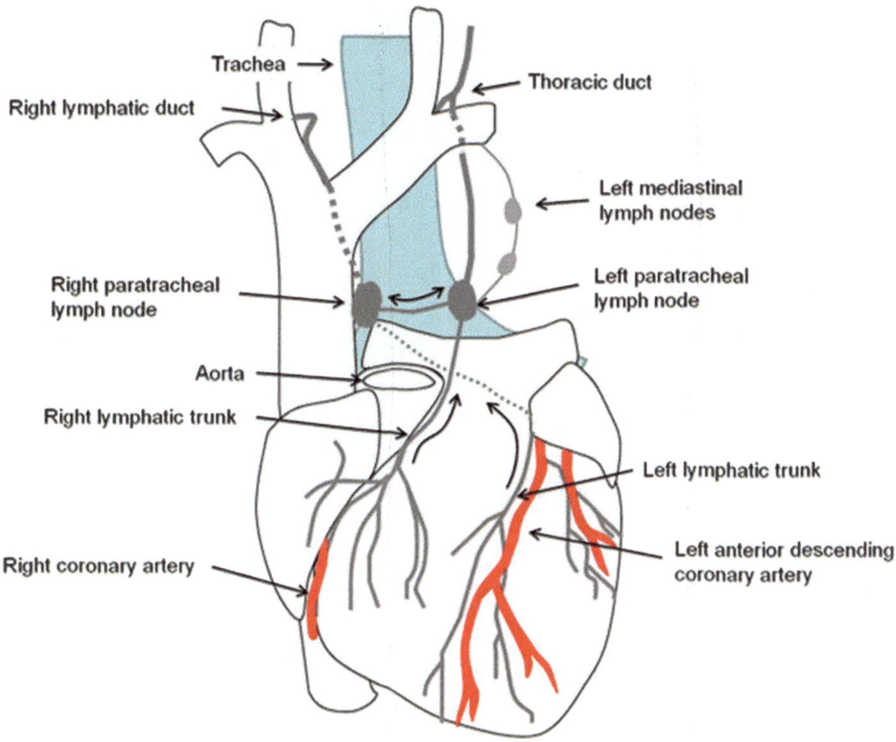

Fig. 6.1 The right lymphatic trunk drains into the left paratracheal nodes and the left anterior mediastinal nodes, and the left lymphatic trunk drains into the right paratracheal nodes. The right and left paratracheal nodes are connected

The cardiac lymphatic system has an important role in myocardial fluid balance after myocardial infarction. Ishikawa et al. [12] reported on cardiac lymphatics in various healing stages after acute myocardial infarction. Newly formed lymphatic vessels first appeared in the early stages of granulation and subsequently increased in the late stages. The lymphatics then remained up to the scar phase of the infracted myocardium. Thus, increased myocardial edema as seen in infarcted myocardium eventually led to increased lymph flow. This discovery suggests that cardiac lymphatics are essential for fluid balance in the infarcted area. These findings were supported in experimental canine myocardial infarction models. Lymphatic flow and protein content increased 3 hours after the occlusion of the left anterior descending coronary artery [13]. Park et al. [14] also demonstrated in an experimental murine model of myocardial infarction that lymphatic vessels were not detected in lesions with areas of necrosis/infarction and new lymphatics were well formed in the peri-infarction areas 2 weeks after myocardial infarction (Fig. 6.2).

Fluid balance maintenance of the pericardial space is another important lymphatic function. Obstruction of venous blood flow and lymphatic outflow can lead to pericardial effusion formation [15]. Pericardial effusion originates from the

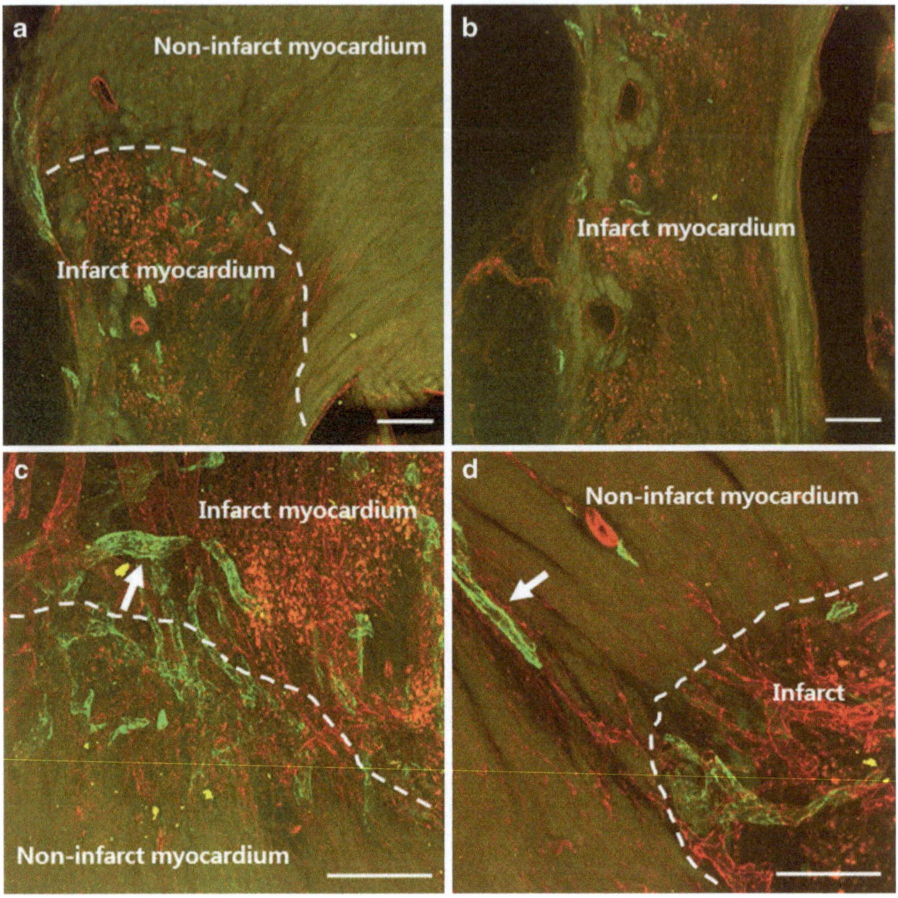

Fig. 6.2 Cross-sectional image of the heart 2 weeks after acute myocardial infarction (*red* fluorescence represents platelet endothelial cell adhesion molecule (PECAM) for blood vessels, and green fluorescence represents lymphatic vessel endothelial receptor 1 (LYVE-1) for lymphatic vessels). (**a**) Lymphatic vessels were rarely seen in the noninfarct area. (**b**) Lymphatic vessels were rarely seen in the infarct area. However, lower panels demonstrate newly formed lymphatic vessels in the peri-infarction area (*arrow*, **c**, **d**). Bars, 100 μm (modified with permission from Park et al. [14])

visceral pericardium, and the degree of pericardial effusion is proportionate to the extent of venous and lymphatic flow interference. This function can be useful in understanding the pathologic mechanisms of pericardial effusion in cases of acute myocarditis or constrictive pericarditis [6].

A third cardiac lymphatic function is its immunologic and anti-inflammatory effect. Intravenous injection of staphylococcal bacteria in dogs with chronically impaired cardiac lymph flow resulted in bacterial endocarditis and myocarditis [16]. Lymphatic impairment may predispose an inflammatory process in the heart valves, accompanied by possible decreased pathogen removal.

The cardiac lymphatic system is also a major route for cancer metastasis [17]. Cancer metastasis is an important cause of lymphatic obstruction of the heart. Another potential yet controversial role for the cardiac lymphatic system is an anti-atherosclerotic function [18]. The development of atherosclerosis in the intramural coronary arteries is prevented by a highly effective pericoronary lymphatics and venous capillary system. There is increased incidence of atherosclerotic lesions in the allogenic transplanted heart because of the failure of the cardiac lymphatic system [19]. A similar atherogenic condition was observed as well in Kawasaki disease with associated atherosclerosis acceleration [20].

Origin of Newly Formed Lymphatic Vessels

The generation of lymphatic vessels in adults was previously believed to be achieved exclusively by a process called lymphangiogenesis, the formation of new lymphatic vessels from preexisting lymphatic vasculature [21–25]. However, emerging evidence has suggested that lymphvasculogenesis may also occur through putative progenitor cells for lymphatic/vascular endothelial precursor cells (LECs). Studies have shown that bone marrow (BM) contains cells with the potential to generate LECs [26–29]. A study demonstrated that BM hematopoietic cells are the source of lymphvasculogenic CD11b+cells and, using in vitro culture methods and in vivo physiological and pathological conditions, further identified a specific cell population among the heterogeneous CD11b+, i.e., pod+CD11b+, cells that can function as lymphatic endothelial progenitor cells (EPC) [30]. Although specific populations of BM cells contribute to lymphvasculogenesis, few BM-derived cells contribute to newly formed lymphatic vessels in the peri-infarction areas [14].

Cardiac Lymphatics in Cardiovascular Diseases

Heart Failure

Because organized myocardial contraction is the major determinant of myocardial lymph flow, myocardial lymphatic flow can be decreased with the impairment of ventricular function in heart failure [31]. This has been demonstrated with canine models of cardioplegic arrest, which resulted in decreased lymphatic drainage and subsequent interstitial myocardial edema. The decreased lymphatic flow was improved significantly following return of sinus rhythm after recovery from cardioplegic arrest. The interruption of cardiac lymphatic flow may play an important role in the pathogenesis of heart failure after heart transplantation [19]. The blocking of cardiac lymph flow can induce myocardial edema and ventricular dysfunction in experimental models [32]. There are several proposed mechanisms for the development of ventricular dysfunction. One possible mechanism is cardiac edema induced by blocking cardiac lymph flow. Myocardial edema could then cause ventricular dysfunction by restricting oxygen supply and inducing ischemia [33]. Another proposed mechanism suggests that blocking cardiac lymph flow may cause an increase in the interstitial concentration of metabolites that have negative inotropic effects [10]. These cardiotoxic metabolites depress ventricular contractility without concomitant increase in myocardial water content.

In a chronic heart failure animal model, cardiac lymph flow increased compared to normal controls, and lymphatic vessels expanded in their size as well [18]. This finding is similar to a study of heart failure patients who also exhibited distended and disturbed lymphatic flow [34]. This suggests that the capacity of the lymphatic system to transport excess capillary filtrate back to the blood stream serves as a major control mechanism resulting in the manifestation of heart failure. Increased interstitial edema of the myocardium reduced myocardial performance and increased collagen levels [35].

Generally, the presence of lymphatics has not been demonstrated in the heart valves. However, there is one report of lymphatics in mitral valves of patients who died with chronic heart failure, where lymphatic vessels were distributed beneath the endocardium on the atrial surface of the posterior cusp of the mitral valve [3]. The presence of cardiac lymphatics in the valves might have resulted from an increased edematous portion of that region.

Myocardial Infarction

After acute occlusion of the coronary arteries, there is a decreased blood supply, increased vascular permeability due to capillary endothelial injury, and removal of fluid from the ischemic myocardium. These events result in interstitial edema of that lesion. Cardiac lymphatic flow was seen to decrease dramatically within the first 30 min after occlusion of the coronary artery [36]. When edema formation occurs within the interstitium of the newly infarcted heart, structural and functional remodeling of the myocardium can occur [37].

In the healing stages of acute myocardial infarction, the cardiac lymphatic system is required to remove myocardial edema. New lymphatic vessels form within the infarcted myocardium and appear to be important for the remodeling process that occurs after myocardial infarction [12].

An increase in the number of lymphatic vessels accompanies acute myocardial infarction, particularly at the edges of the necrotic areas. Although a local increase in lymphatic vessel density was found in both acute and chronic ischemia, the increase was limited to certain pathological areas, such as necrotic edges, scars, and reactive pericarditis. Ischemia was accompanied by the formation of new vessels as both blood and lymphatic vessels featured dilatation, branching, and sprouting. Over the past decade, intensive efforts have been undertaken to develop therapeutic strategies to promote revascularization in ischemic tissues. Proangiogenic therapy uses vascular growth factors to induce therapeutic vascular growth [38]. Greater understanding of the cardiac lymphatic system should help in the design of better proangiogenic therapeutic approaches and raises the possibility of influencing both preexisting and newly formed lymphatics in normal and diseased human hearts, particularly in valve and advanced coronary diseases [38]. Cell therapy using BM-derived EPC transplantation after myocardial infarction usually showed beneficial effects in left ventricular systolic performance as well as decreased left ventricular fibrosis. Cell therapy did not increase lymphangiogenetic cytokines or lymphatic vessel number compared to a control group in a murine model, which means that injected EPCs enhanced the healing process after myocardial infarction, reduced interstitial edema, and decreased the formation of lymphatic vessels (Fig. 6.3) [14].

Valvular Heart Diseases

Some investigators have shown endocardial deposition of fibrous and elastic tissues, as well as thickening and opacity of mitral valve leaflets in a chronic myocardial injury model by blocking cardiac lymph flow [39]. They proposed that ventricular subendocardial hemorrhage occurred after lymphatic impairment had led to chronic fibrosis of the valves. Another study observed chronic obstruction of cardiac lymphatic flow in thickened atrioventricular valves [40]. These changes were considered to be caused by the accumulation of amorphous material, the major components of which were hyaluronic acid and chondroitin sulfates. Chronic lymphatic obstruction can partly account for the development of myxoid accumulation in the valve leaflets.

Myocardial Dysfunction and Allograft Vasculopathy in the Transplanted Heart

Cardiac lymphatic interruption during heart transplantation is a possible explanation of myocardial dysfunction and allograft vasculopathy in the transplanted heart [19]. Like acute disruption of cardiac lymphatics, subsequent myocardial edema can reduce myocardial contractility. Also, impaired lymphatic flow and interstitial edema can increase susceptibility to endocardial or myocardial infection in immune-suppressed heart transplant patients [16]. Coronary vasculopathy is another problem in heart transplant patients. Cardiac allograft vasculopathy may be a long-term consequence of lymphatic stasis [41].

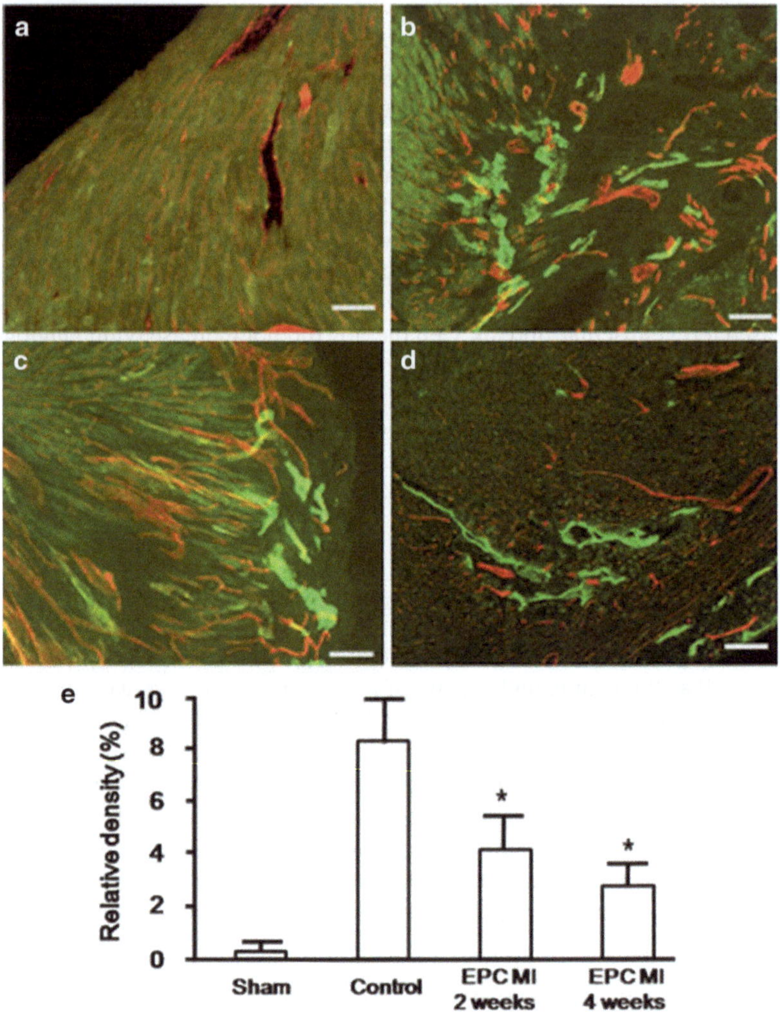

Fig. 6.3 Cross-sectional images of the heart 2 and 4 weeks after acute myocardial infarction (AMI, *red* fluorescence represents platelet endothelial cell adhesion molecule (PECAM) for blood vessels, and *green* fluorescence represents lymphatic vessel endothelial receptor 1 (LYVE-1) for lymphatic vessels). (**a**) Lymphatic vessels were seldom found in the sham operation group. (**b**) Newly formed lymphatic vessels were developed in the peri-infarction area in the PBS group (control). However, the endothelial progenitor cell (EPC) treatment group shows a lesser degree of lymphatics at 2 (**c**) and 4 weeks (**d**) after AMI. Relative density of lymphatic vessels was significantly lower in the EPC treatment group (**e**). Bars, 100 μm, $*P<0.05$ (reproduced with permission from Park et al. [14])

Therapeutic Challenges with Cardiac Lymphatics

In Heart Failure

Because increased myocardial edema is related to increased fibrosis of the myocardium, rapid resolution of myocardial edema can reduce subsequent fibrosis [35]. Transient use of inotropic agents can be used in this case to increase cardiac lymphatic drainage. In canine cardioplegia models, enhancement of cardiac contractility through the intravenous infusion of dobutamine ultimately increased lymphatic return [42]. The use of dobutamine was associated with decreased extent of myocardial edema compared to controls. Also, the use of potent diuretics can decrease the extent of myocardial fibrosis in heart failure patients by reducing myocardial edema at a quicker rate [43].

In other studies, Witte et al. [34] studied enhanced lymphatic drainage in patients with severe intractable congestive heart failure after cannulation of the thoracic duct. They found that symptoms of circulatory congestion were dramatically relieved after venting the distended thoracic duct. However, the effect of lymphatic drainage via the thoracic duct is relatively limited.

In Myocardial Ischemia and Infarction

Because myocardial infarction is one of the major causes of heart failure, both prevention of acute myocardial infarction and prompt management enhancing favorable ventricular remodeling are needed in order to reduce heart failure. Improving cardiac lymphatic flow may have beneficial effects on cardiac contractility following myocardial infarction [44, 45]. Park et al. [14] reported in a murine model that injection of EPC into the peri-infarct area was associated with reduced lymphatic vessel density in that region and favorable ventricular remodeling. This favorable effect of EPC can be correlated with the reduction of myocardial edema in the region. Other studies demonstrated that the use of hyaluronidase maintained lymphatic flow after induction of myocardial infarction and also decreased the extent of myocardial infarction [36, 46]. In addition, several randomized controlled trials with intravenous hyaluronidase infusion have demonstrated mortality benefit in the treatment group [47–49]. However, the necessity of administration within 6 h of symptom onset and the modest benefit prohibited the widespread clinical use of this particular agent [47–49].

Early management enhancing myocardial healing, after restoration of blood flow in acute myocardial infarction, can potentially reduce myocardial edema and decrease subsequent myocardial fibrosis. Augmentation of lymphatic flow by using hyaluronidase is one of the interventions that could be used to improve clinical outcomes [50, 51]. During the healing stages following myocardial infarction, hyaluronidase facilitates lymphatic drainage from the infarcted area [50]. This favorable effect of hyaluronidase would reduce the vulnerability of the ischemic myocardium and thus dampen the effects of toxic metabolites that mediate reperfusion injury of the lesion.

In Heart Transplantation

Hyaluronidase use after heart transplantation reduced myocardial interstitial edema in the recipient [52]. Moreover, administration of this agent during cardioplegic arrest was seen to increase cardiac lymphatic drainage and reduce myocardial edema. The favorable effects of hyaluronidase in this case are therefore similar to ischemia or infarct models.

Future Perspectives

Cardiac lymphatics can contribute to various disease processes of the heart. Further investigation may elucidate the pathogenic mechanism of fibrotic diseases in heart valves and inflammatory pericardial diseases. Lymph flow impairment caused by lymphatic disruption or flow resistance may occur in many disease processes such as myocardial infarction. Because myocardial infarction is the most common cause of heart failure, prevention of adverse ventricular remodeling with functional lymphatic regeneration (lymphangiogenesis) will improve cardiac lymph flow and can reduce the number of heart failure patients. Modulation of the cardiac lymphatic system can also be used to reduce the incidence of chronic allograft rejection after heart transplantation.

Conclusion

Although the cardiac lymphatic system has several important roles in the normal and diseased heart, its importance has been overlooked by many physicians. For example, there is increased risk of edematous myocardium in heart failure patients, and persistent myocardial edema may result in myocardial fibrosis and further decreasing of systolic function. Early resolution of myocardial edema can reduce the amount of myocardial fibrosis and prevent subsequent myocardial functional loss. Therefore physicians should be aware of fluid balance of the heart and the importance of cardiac lymphatics during their practice.

References

1. Mehlhorn U, Geissler HJ, Laine GA, Allen SJ (2001) Role of the cardiac lymph system in myocardial fluid balance. Eur J Cardiothorac Surg 20:424–427
2. Shimada T, Morita T, Oya M, Kitamura H (1990) Morphological studies of the cardiac lymphatic system. Arch Histol Cytol 53(Suppl):115–126
3. Johnson RA, Blake TM (1966) Lymphatics of the heart. Circulation 33:137–142

4. Riquet M, Hidden G (1991) Lymphatic drainage of the left atrium and ventricle of the heart. Surg Radiol Anat 13:238–240
5. Riquet M, Hidden G (1991) Lymphatic drainage of the right atrium and ventricle of the heart. Surg Radiol Anat 13:235–237
6. Miller AJ, DeBoer A, Pick R, Van Pelt L, Palmer AS, Huber MP (1988) The lymphatic drainage of the pericardial space in the dog. Lymphology 21:227–233
7. Leeds SE, Uhley HN, Meister RB, McCormack KR (1977) Lymphatic pathways and rate of absorption of 131I-albumin from pericardium of dogs. Lymphology 10:166–172
8. Miller AJ, Pick R, Katz LN (1964) The importance of the lymphatics of the mammalian heart: experimental observations and some speculations. Circulation 29(Suppl):485–487
9. Sun SC, Lie JT (1977) Cardiac lymphatic obstruction: ultrastructure of acute-phase myocardial injury in dogs. Mayo Clin Proc 52:785–792
10. Ludwig LL, Schertel ER, Pratt JW et al (1997) Impairment of left ventricular function by acute cardiac lymphatic obstruction. Cardiovasc Res 33:164–171
11. Wang YL, Wang XH, Liu YL, Kong XQ, Wang LX (2009) Cardiac lymphatic obstruction impairs left ventricular function and increases plasma endothelin-1 and angiotensin II in rabbits. Lymphology 42:182–187
12. Ishikawa Y, Akishima-Fukasawa Y, Ito K et al (2007) Lymphangiogenesis in myocardial remodelling after infarction. Histopathology 51:345–353
13. Feola M, Glick G (1975) Cardiac lymph flow and composition in acute myocardial ischemia in dogs. Am J Physiol 229:44–48
14. Park JH, Yoon JY, Ko SM et al (2011) Endothelial progenitor cell transplantation decreases lymphangiogenesis and adverse myocardial remodeling in a mouse model of acute myocardial infarction. Exp Mol Med 43:479–485
15. Miller AJ, Pick R, Johnson PJ (1971) The production of acute pericardial effusion: the effects of various degrees of interference with venous blood and lymph drainage from the heart muscle in the dog. Am J Cardiol 28:463–466
16. Miller AJ, Pick R, Kline IK, Katz LN (1964) The susceptibility of dogs with chronic impairment of cardiac lymph flow to staphylococcal vagular endocarditis. Circulation 30:417–424
17. Kondo T, Kitazawa R, Kawata E, Mori K, Kitazawa S (2009) Diffuse cardiac lymphatic involvement by metastatic neuroendocrine carcinoma mimicking hypertrophic cardiomyopathy: a case report. Cases J 2:9127
18. Uhley HN, Leeds SE, Sampson JJ, Friedman M (1969) The cardiac lymphatics in experimental chronic congestive heart failure. Proc Soc Exp Biol Med 131:379–381
19. Kong XQ, Wang LX, Kong DG (2007) Cardiac lymphatic interruption is a major cause for allograft failure after cardiac transplantation. Lymphat Res Biol 5:45–47
20. Noto N, Okada T, Abe Y et al (2012) Characteristics of earlier atherosclerotic involvement in adolescent patients with Kawasaki disease and coronary artery lesions: significance of gray scale median on B-mode ultrasound. Atherosclerosis 222:106–109
21. He Y, Rajantie I, Ilmonen M et al (2004) Preexisting lymphatic endothelium but not endothelial progenitor cells are essential for tumor lymphangiogenesis and lymphatic metastasis. Cancer Res 64:3737–3740
22. Karpanen T, Egeblad M, Karkkainen MJ et al (2001) Vascular endothelial growth factor C promotes tumor lymphangiogenesis and intralymphatic tumor growth. Cancer Res 61:1786–1790
23. Karkkainen MJ, Haiko P, Sainio K et al (2004) Vascular endothelial growth factor C is required for sprouting of the first lymphatic vessels from embryonic veins. Nat Immunol 5:74–80
24. Nagy JA, Vasile E, Feng D et al (2002) Vascular permeability factor/vascular endothelial growth factor induces lymphangiogenesis as well as angiogenesis. J Exp Med 196:1497–1506
25. Cao R, Bjorndahl MA, Gallego MI et al (2006) Hepatocyte growth factor is a lymphangiogenic factor with an indirect mechanism of action. Blood 107:3531–3536
26. Salven P, Mustjoki S, Alitalo R, Alitalo K, Rafii S (2003) VEGFR-3 and CD133 identify a population of CD34+ lymphatic/vascular endothelial precursor cells. Blood 101:168–172

27. Maruyama K, Ii M, Cursiefen C et al (2005) Inflammation-induced lymphangiogenesis in the cornea arises from CD11b-positive macrophages. J Clin Invest 115:2363–2372
28. Religa P, Cao R, Bjorndahl M, Zhou Z, Zhu Z, Cao Y (2005) Presence of bone marrow-derived circulating progenitor endothelial cells in the newly formed lymphatic vessels. Blood 106:4184–4190
29. Kerjaschki D, Huttary N, Raab I et al (2006) Lymphatic endothelial progenitor cells contribute to de novo lymphangiogenesis in human renal transplants. Nat Med 12:230–234
30. Lee JY, Park C, Cho YP et al (2010) Podoplanin-expressing cells derived from bone marrow play a crucial role in postnatal lymphatic neovascularization. Circulation 122:1413–1425
31. Mehlhorn U, Davis KL, Burke EJ, Adams D, Laine GA, Allen SJ (1995) Impact of cardiopulmonary bypass and cardioplegic arrest on myocardial lymphatic function. Am J Physiol 268:H178–H183
32. Ullal SR, Kluge TH, Gerbode F (1972) Functional and pathologic changes in the heart following chronic cardiac lymphatic obstruction. Surgery 71:328–334
33. Gloviczki P, Solti F, Szlavy L, Jellinek H (1983) Ultrastructural and electrophysiologic changes of experimental acute cardiac lymphostasis. Lymphology 16:185–192
34. Witte MH, Dumont AE, Clauss RH, Rader B, Levine N, Breed ES (1969) Lymph circulation in congestive heart failure: effect of external thoracic duct drainage. Circulation 39:723–733
35. Laine GA, Allen SJ (1991) Left ventricular myocardial edema. Lymph flow, interstitial fibrosis, and cardiac function. Circ Res 68:1713–1721
36. Szlavy L, Koster K, de Courten A, Hollenberg NK (1987) Early disappearance of lymphatics draining ischemic myocardium in the dog. Angiology 38:73–84
37. Santos AC, de Lima JJ, Botelho MF et al (1998) Cardiac lymphatic dynamics after ischemia and reperfusion–experimental model. Nucl Med Biol 25:685–688
38. Kholova I, Dragneva G, Cermakova P et al (2011) Lymphatic vasculature is increased in heart valves, ischaemic and inflamed hearts and in cholesterol-rich and calcified atherosclerotic lesions. Eur J Clin Invest 41:487–497
39. Miller AJ, Pick R, Katz LN (1960) Ventricular endomyocardial pathology produced by chronic cardiac lymphatic obstruction in the dog. Circ Res 8:941–947
40. Symbas PN, Schlant RC, Gravanis MB, Shepherd RL (1969) Pathologic and functional effects on the heart following interruption of the cardiac lymph drainage. J Thorac Cardiovasc Surg 57:577–584
41. Weis M, von Scheidt W (1997) Cardiac allograft vasculopathy: a review. Circulation 96:2069–2077
42. Allen SJ, Geissler HJ, Davis KL et al (1997) Augmenting cardiac contractility hastens myocardial edema resolution after cardiopulmonary bypass and cardioplegic arrest. Anesth Analg 85:987–992
43. Lopez B, Querejeta R, Gonzalez A, Sanchez E, Larman M, Diez J (2004) Effects of loop diuretics on myocardial fibrosis and collagen type I turnover in chronic heart failure. J Am Coll Cardiol 43:2028–2035
44. Cui Y (2010) The role of lymphatic vessels in the heart. Pathophysiology 17:307–314
45. Cui Y (2010) Impact of lymphatic vessels on the heart. Thorac Cardiovasc Surg 58:1–7
46. Szlavy L, Adams DF, Hollenberg NK, Abrams HL (1980) Cardiac lymph and lymphatics in normal and infarcted myocardium. Am Heart J 100:323–331
47. Cairns JA, Holder DA, Tanser P, Missirlis E (1982) Intravenous hyaluronidase therapy for myocardial infarction in man: double-blind trial to assess infarct size limitation. Circulation 65:764–771
48. Roberts R, Braunwald E, Muller JE et al (1988) Effect of hyaluronidase on mortality and morbidity in patients with early peaking of plasma creatine kinase MB and non-transmural ischaemia. Multicentre investigation for the limitation of infarct size (MILIS). Br Heart J 60:290–298
49. Saltissi S, Robinson PS, Coltart DJ, Webb-Peploe MM, Croft DN (1982) Effects of early administration of a highly purified hyaluronidase preparation (GL enzyme) on myocardial infarct size. Lancet 1:867–871

50. Repa I, Garnic JD, Hollenberg NK (1990) Myocardial infarction treated with two lymphagogues, calcium dobesilate (CLS 2210) and hyaluronidase: a coded, placebo-controlled animal study. J Cardiovasc Pharmacol 16:286–291
51. Szlavy L, Repa I, Lengyel I, Lamboy L (1990) Calcium dobesilate (CLS 2210) protects the myocardium in early acute myocardial infarction: a preliminary randomized, double-blind, placebo-controlled study of its effects on biochemical markers. J Cardiovasc Pharmacol 15:89–95
52. Johnsson C, Hallgren R, Elvin A, Gerdin B, Tufveson G (1999) Hyaluronidase ameliorates rejection-induced edema. Transpl Int 12:235–243

Chapter 7
Imaging the Lymphatic System in Heart Transplantation and Its Immunological Implications

Kathryn Brown, Lindsey A. Edwards, and Wilson Wong

Abstract Cardiac transplantation is an important form of treatment for patients with end-stage heart failure. Rejection by the immune system of the recipient is a major cause of graft loss. The lymphatic system plays a crucial role in the immune response, but has not received much attention. Donor blood vessels are joined to those of the recipient during the transplant operation; however, due to their small size, donor lymphatic vessels are not reconnected. For this same reason the imaging of lymphatics, and therefore the understanding of this system, has been hampered. In recent years great strides have been made towards developing new imaging modalities that allow the lymphatic system to be imaged in a non-invasive or semi-invasive manner. Not only does this imaging provide new insights into the lymphatic system, it also presents the possibility of using lymphatic imaging to diagnose and monitor diseases and their treatment. In the case of heart transplantation, which will be the focus of this chapter, the lymphatic system has been shown to play a role in the alloimmune response. Imaging may be used as an extra tool for patient monitoring or as a research tool to develop new treatment protocols for recipients of cardiac transplants.

Keywords Lymphatic system • Imaging • Heart • Transplantation

Abbreviations

CEUS	Contrast-enhanced ultrasound
CT	Computed tomography
DC	Dendritic cell

K. Brown (✉) • L.A. Edwards • W. Wong
MRC Center for Transplantation, King's College London School of Medicine at Guy's, King's and St. Thomas' Hospitals UK, Great Maze Pond, London SE1 9RT, UK
e-mail: kathryn.brown@kcl.ac.uk

G. Karunamuni (ed.), *The Cardiac Lymphatic System: An Overview*,
DOI 10.1007/978-1-4614-6774-8_7,

ICAM-1	Intercellular adhesion molecule 1
ICG	Indocyanine green
LYVE-1	Lymphatic vessel endothelial receptor-1
MRI	Magnetic resonance imaging
NIR	Near-infrared
PET	Positron emission tomography
Prox1	Prospero homeobox protein-1
Qdots	Quantum dots
SPECT	Single-photon emission computed tomography
SPIO	Superparamagnetic iron oxide particles
Tc-99m	Technetium-99m
VEGFR3	Vascular endothelial growth factor receptor 3

Introduction

Heart transplantation is the best form of treatment for patients with end-stage heart failure. Although 1-year survival rates are favourable, at over 85 %, the average half-life of a heart graft is only 10 years. Chronic rejection remains the major cause of graft failure. Following heart transplantation, the injury inevitably caused by harvesting of the organ and the transplant surgery leads to inflammation within the graft. The resulting cytokine milieu activates cells present within the graft, particularly donor-derived antigen-presenting cells. The activation of these cells causes their migration out of the graft via the lymphatic vessels and blood into the secondary lymphoid organs of the recipient, where they can present donor antigens to recipient T cells. Recipient T cells can then traffic into the heart graft and elicit an immune response, with cytotoxic T cells killing target cells directly and helper T cells promoting macrophage and B cell responses. The initial inflammatory conditions within the graft also recruit recipient cells, including neutrophils and macrophages, which damage the graft, and antigen-presenting cells, which can phagocytose and present fragments of donor antigens to recipient T cells, thus perpetuating the alloresponse.

Despite the importance of the lymphatic system in leukocyte trafficking, its pathophysiology following vascularised organ transplantation has received little attention, unlike the situation in corneal transplants. Corneal grafts are immunoprivileged, due to the lack of both lymphatic and blood supplies. As a result, donor passenger leukocytes, the main source of allostimulation for the immune system of the host, cannot leave the donor graft, resulting in immune deviation. This phenomenon is well characterised (summarised in [1]).

The situation is more complicated in vascularised organs where donor passenger leukocytes can exit the graft via either blood or lymphatics. Remodelling of lymphatics post-transplantation can be divided into (1) reconnection of donor lymphatics to that of the recipient and (2) de novo neogenesis of lymphatic vessels within the donor organ as a result of inflammation [2], such as infiltration by macrophages which express cell surface molecules associated with lymphangiogenesis: LYVE-1, Prox-1, VEGFR-3 and VEGF-C [3].

In the absence of intact lymphatics after transplantation, donor dendritic cells (DC) may traffic to the spleen via reverse transmigration into the bloodstream, as shown for plasmacytoid DC after heart transplantation [4]. Alternatively, DC may leave the donor organ via the open severed ends of the lymphatic vessels into the immediate surrounding area. They will then be taken up by the lymphatic capillaries in the local vicinity and traffic to the local draining lymph nodes. It is likely that the efficiency in terms of speed and number of cells transported would be reduced compared to that of intact lymphatic vessels. Only a limited number of studies have been carried out on this important topic, providing circumstantial evidence only and with conflicting results, with some suggesting a beneficial role for lymphatics while others suggest the opposite.

Evidence for lymphatics being beneficial in transplantation: It has been shown that human cardiac allografts that have undergone rejection had a lower density of VEGFR-3+ lymphatics compared to patients with only moderate rejection [5]. In an experimental model of lung transplantation in dogs, lack of lymphatic drainage was associated with rejected organs [6]. Furthermore, in rat cardiac allografts, a decrease in lymphatic density has been observed in the inner myocardium post-transplant [7]. It has been proposed that the decrease in density of lymphatics may contribute to rejection through the retention of graft-destructive cells, as well as the accumulation of cytokines that perpetuate the local immune response. In support of this theory, it has been shown that mechanical interruption of lymphatic drainage induces expression of a panel of genes involved in acute inflammation, complement activation and B cell humoral immune responses [8]. Certainly, lymphatic vessels in human kidney transplants are associated with lymphocytic infiltrates [9]. Although the density of lymphatic vessels was not associated with acute or chronic rejection, grafts with lymphatic vessels within infiltrates had better graft function one year after transplantation compared to those without such vessels [10]. Put together, these data provide circumstantial evidence that increased lymphatic vessel density within transplanted vascularised organs is associated with better graft outcome. This suggests that pharmacological modulation of lymphangiogenesis may be a possible strategy to improve graft outcome.

Evidence for lymphatics being detrimental in transplantation: Conversely, it has been suggested that lymphatics may also play the opposite role and decrease graft survival. Increased lymphatic drainage from donor organs may facilitate antigen-presenting cell trafficking to secondary lymphoid organs, promoting alloimmunity [2]. In our laboratory, using a mouse kidney model of tolerance, the density of lymphatic vessels increased with time; however, their presence within the tertiary lymphoid organs that formed within the allografts was associated with reduced graft function [11]. In a study of human renal transplant biopsy material, transplant rejection was observed to be caused by a lymphocyte-rich inflammatory infiltrate that attacked cortical tubules and endothelial cells. This corresponded with a >50-fold increase of lymphatic vessel density. Numerous chemokine receptor CCR7(+) cells within the infiltrates seemed to be attracted by secondary lymphatic chemokine (SLC: CCL21) that is produced and released by lymphatic endothelial cells in a complex with the lymphatic marker podoplanin. The authors speculated that lymphatic neoangiogenesis not only contributes to the export of the rejection infiltrate but is also involved in the maintenance of alloreactive immune response in renal transplants and thus provides a novel therapeutic target [9]. Indeed, it has been demonstrated in islet allografts that interfering with lymphatic function leads to inhibition of lymphangiogenesis and prolonged or indefinite allograft survival [12].

It is not known whether an enhanced and more physiological connection between donor and recipient lymphatic vessels, through which DC can traffic, may result in a stronger or weaker immune response. Evidence to date for and against lymphatics being beneficial in transplantation is patchy and mainly observational. Whether changes in lymphatic density detected on histological studies are cause or effect of the immune process is unknown. More importantly, higher density of lymphatic vessels within donor organs does not necessarily mean better lymphatic flow or more efficient transport of immune cells such as DC and T cells. The effects of transplant rejection itself on the structure and function of lymphatic vessels remains unclear.

In addition to heart transplantation, imaging of the lymphatics will be of benefit in many cardiac diseases and thus will have a wide range of applications. Ischaemia/reperfusion injury affects lymphatic flow from the heart, and this change is seen before myocardial necrosis is visible. Such changes in lymphatic drainage could be an early marker of heart disease and allow an early diagnosis [13, 14]. It has also been hypothesised that ineffective lymphatic drainage from the epicardial coronary arteries contributes to the development of atherosclerosis and that the accelerated coronary atherosclerosis seen in heart transplant recipients is a result of the interrupted lymphatic drainage after surgery [15]. Imaging this system may therefore prove to be a powerful research tool and clinical investigation modality to benefit all cardiac patients, not just transplant recipients.

The difficulty of imaging lymphatics, due to their small size, has contributed to the relative lack of information about this system compared with the blood circulation. Recent developments in this area have already led to advances in our understanding of the lymphatic system, and this should continue as the imaging techniques and tools are improved and refined.

Imaging of the lymphatic system has primarily been driven in the field of cancer research—the techniques have been developed and utilised to detect metastatic tumours and to identify the sentinel (draining) lymph node of a tumour in order to perform a biopsy. However, these techniques can be used to image lymphatics under any conditions, including after heart transplantation.

Lymphatic imaging techniques typically involve the injection into tissues of agents, the size and characteristics of which result in uptake by the lymphatic system. Molecules of around 10 nm are thought to be ideal for imaging lymphatics, as they enter the lymphatic capillaries and highlight the lymphatic vessels and the draining lymph nodes. Molecules smaller than 6 nm also enter the lymphatic system but diffuse out quickly leading to a less defined image [16]. Larger molecules, up to 100 nm, image the lymphatic system through their phagocytosis by tissue resident macrophages, which then carry them to the draining lymph nodes, while molecules bigger than 100 nm cannot leave the interstitium [17] and are therefore not suitable.

Lymphatics of the heart are not routinely imaged in clinical practice. Only isolated examples have been described. However, numerous techniques have been

Table 7.1 Advantages and disadvantages of the techniques used to image the lymphatic system

Imaging technique	Pros	Cons	Refs
Blue dye injection	No ionising radiation Low cost Widely available	Invasive Can only image area of interest	[18–20]
NIR imaging	Non-invasive No ionising radiation Quantitative	Very poor depth penetration	[21–23]
Qdots	Non-invasive No ionising radiation Quantitative Low cost Widely available	Poor depth penetration Potentially toxic	[24–26]
Planar lymphoscintigraphy	Non-invasive Quantitative	Ionising radiation Low resolution Poor localisation	[27–30]
SPECT	Non-invasive Quantitative Good resolution	Ionising radiation Poor localisation High cost Not widely available	[31, 32]
CT	Non-invasive Quantitative Good localisation	Ionising radiation Iopamidol quickly leaves LNs High cost Not widely available	[33–35]
SPECT/CT	Non-invasive Quantitative Good resolution Good localisation	Ionising radiation High cost Not widely available	[48–50]
MRI	Non-invasive Quantitative No ionising radiation Excellent resolution Good localisation	SPIOs traffic slowly Long half-life of agents High cost Not widely available	[34, 36–46]
CEUS	Non-invasive Quantitative No ionising radiation Good contrast Low cost Widely available	Mechanism unclear Microbubbles not ideal agents	[47, 59]

adopted in the experimental setting to study various aspects of lymphatic drainage of the heart. Here, we will summarise the main techniques and their applications.

Two main techniques are currently used to image the lymphatic system in the clinic. These are dye injection and lymphoscintigraphy, which will be discussed below. Both of these methods have drawbacks and therefore new techniques and imaging agents are being developed. These new methods will also be described below, with a summary of the data in Table 7.1.

Methods of Lymphatic Imaging

Dye Injection

The first method used to detect lymphatics in living animals, following on from the injection of air or mercury into tissues, was injection of a blue dye (Evans blue, isosulfan blue, India ink or patent blue V) into tissues, including the endocardium, myocardium and epicardium. These dyes selectively enter the lymphatic vessels, allowing them to be visualised macroscopically. They also highlight the draining lymph node, which has made this an important technique in sentinel node biopsy of cancer patients. This method was used to describe the lymphatic drainage of the canine heart [18], results that were subsequently confirmed in humans using injection of blue dye into both cadavers [19] and live hearts during surgery [20]. As such it remains one of the main ways to visualise lymphatics in the clinic. The main drawback of this technique in terms of clinical use is its invasive nature, with dissection needed to reveal the lymphatics and lymph nodes before the dye can be seen. This also means that lymphatic drainage into any unexpected, and thus un-dissected, areas will go undetected.

Optical Imaging

Optical imaging is a refinement of the technique of blue dye injection, by using fluorescent molecules to detect the lymphatics instead. This has the advantage of being able to use multiple colours to define lymphatics draining from separate regions, and also the fluorophores can be detected through tissue.

Near-Infrared Fluorophores

Near-infrared (NIR) fluorophores (reviewed in [21]) are preferred due to reduced background when imaging in this part of the spectrum. Indocyanine green (ICG) has already been approved for use in humans to image lymphatics, and other small NIR fluorophores such as Alexa 700, Cy7 and Cy5.5 can be used after conjugation to macromolecules. Subsequent to injection and uptake into the lymphatics, these fluorophores are then detected by an NIR camera.

ICG has even been successfully used to quantify lymphatic flow in mouse [22] and rat [23] models. Unfortunately, the brightness of the fluorophores means that only lymphatics just below the surface of the skin (around 1 cm penetration) can be observed, and therefore these molecules are probably unsuitable for cardiac lymphatic imaging.

Quantum Dots

Quantum dots (Qdots) are semiconducting nanoparticles, with far higher fluorescence intensity than traditional NIR fluorophores [24]. Like the traditional NIR fluorophores, multiple molecules with a range of colours can be used concurrently, but unlike these fluorophores, Qdots emit fluorescence in the visible spectrum, bypassing the need for an NIR camera and image processing for detection [25]. The brightness of Qdots as compared with NIR fluorophores also increases the possible depth penetration, up to 2 cm below the skin surface [26]. Further developments along this pathway may possibly allow deeper penetration and enable these techniques to be used for cardiac lymphatic imaging.

Lymphoscintigraphy

Planar Lymphoscintigraphy

Another technique widely used in the clinic, and without the need for invasive surgery for the detection of lymphatics, is lymphoscintigraphy [27]. This involves the injection of a gamma-emitting radioactive tracer into tissues. The most widely used tracers are technetium-99m (Tc-99m)-labelled human serum albumin, sulfur colloid or dextran. These molecules are selectively taken up by the lymphatics and travel along these vessels into the draining lymph nodes. The advantage of this technique is that the lymphatics can be imaged non-invasively, by using a gamma camera to detect the radioactivity. Lymphoscintigraphy has been used successfully to image the lymphatics of the heart in animals [28]. In the case of sentinel lymph node biopsy in the clinic, the images obtained with the gamma camera are used to determine the site of incision, and a gamma probe is then used to find the lymph node, often with the aid of blue dye injection to directly visualise the lymph node [29]. A further benefit of lymphoscintigraphy is the ability to quantify the radioactivity within the lymphatics and lymph nodes from the gamma camera images. This allows the technique to be used to analyse the severity of lymphedema in a non-invasive manner, and it is now the main method of diagnosing and quantifying this condition [30]. The downside of lymphoscintigraphy is the exposure of the patient, and in cases where surgery is involved, the surgeon as well, to ionising radiation. Another problem is the low resolution of the imaging.

Single-Photon Emission Computed Tomography

Single-photon emission computed tomography (SPECT) imaging was developed in the 1970s as an improved tool for lymphoscintigraphy. Instead of the original planar images taken with a gamma camera, SPECT uses a rotating gamma camera that takes 2D images at different angles which are reconstructed to form a 3D image [31]. The development of SPECT has improved the technique of lymphoscintigraphy significantly, due to its increased contrast and resolution, and it also allows more precise anatomical localisation of radioactivity. This improvement has been demonstrated by many comparative studies, where SPECT has been more effective than planar lymphoscintigraphy in identifying sentinel lymph nodes prior to biopsy [32].

Computed Tomography

Computed tomography (CT) involves digital geometry processing utilising computer-processed X-ray images taken around a single axis of rotation to generate a 3D image of the inside of an object [33]. CT has been used for many years for malignant lymph node imaging. The excellent anatomical information obtained using this technique allows the experienced user to distinguish malignant from benign lymph nodes by their increased size [34]. However, the unreliability of this technique led to the use of contrast agents in CT to detect lymphatics. Iopamidol is a water-soluble iodine contrast agent which images lymphatic drainage after interstitial injection [35]. However the low molecular weight of this agent means that it quickly diffuses out of the lymph nodes, allowing only a short time for the imaging of the lymphatic system. This also reduces the value of this technique for quantifying lymphatic flow.

Magnetic Resonance Imaging

Like CT, magnetic resonance imaging (MRI) was previously used to detect malignant lymph nodes mainly through size differences [34]. Also like CT, contrast agents have been developed to image lymphatics specifically using MRI.

Like SPECT, MRI provides cross-sectional images that can be reconstructed into 3D images allowing exact visualisation of lymphatics. MRI also has advantages over SPECT—no ionising radiation is involved, and the resolution obtained is far greater. Body tissue contains water and hence protons (H^+ ions), which become aligned inside the magnetic field of an MRI scanner. A radio frequency current is briefly turned on, producing a varying electromagnetic field. This electromagnetic field has just the right frequency, known as the resonance frequency, to be absorbed and flip the spin of the protons in the magnetic field. After the electromagnetic field is turned off, the spins of the protons become realigned. During this relaxation, a radio frequency signal (electromagnetic radiation in the RF range) is generated, which can be measured with receiver coils. Information about the origin of the signal in 3D space can be learned by applying additional magnetic fields during the scan. Protons in different tissues return to their equilibrium state, producing a signal, at different relaxation rates. T1 is the spin–lattice relaxation time which relates to the recovery of the magnetisation after a resonance frequency pulse, i.e. the time taken so that another pulse can be given and a signal will register. T2 is the spin–spin relaxation time which relates to the decay of magnetisation after resonance frequency pulse, i.e. the time the signal will last after giving a resonance frequency pulse. MRI contrast agents alter the relaxation times of atoms within body tissues and thus alter the radio frequency signal generated and in turn the contrast [36, 37].

There are two basic techniques for the imaging of lymphatics through MRI. The first involves the intravenous or interstitial injection of gadolinium-based contrast agents [38]. These then enter the lymphatics and traffic to the lymph nodes. Gadolinium causes T1 shortening which results in increased signal in MRI scans. The second method to image the lymphatic system uses a different type of MRI contrast agent which is phagocytosed by macrophages, the most common of which are superparamagnetic iron oxide particles (SPIO) [39]. After injection, SPIOs traffic to lymph nodes and are taken up by macrophages, causing accumulation of the contrast agent. This results in a hypointense signal on T2-weighted images. This is not ideal for imaging, and so liposome-based agents containing gadolinium have now been developed to allow T1 shortening and an increase in MR signal in areas of liposome accumulation, which is more easily distinguishable on scans [40].

MRI has been used to image the lymphatic system, and the resolution is high enough to allow visualisation of lymphatic vessels in mice [41]. MRI has also been used following heart transplantation in small animal models [42–46]. In these cases the contrast agents were injected intravenously, and the heart graft was imaged, showing accumulation of the agent in macrophages within the graft. Injection of the contrast agent into the heart graft itself should result in accumulation in the draining lymph nodes and may also allow visualisation of the lymphatics draining from the heart.

Contrast-Enhanced Ultrasound

Ultrasound has also been used to image lymphatics and lymph nodes. The most common contrast agents are microbubbles—gas-filled shells which reflect ultrasound waves [47]. The signal from the microbubbles is very distinct and therefore easily detectable. Microbubbles, when injected into tissues, enter the lymphatics and traffic to the draining lymph nodes, although it is unclear exactly how this trafficking takes place.

Combination of Different Imaging Techniques

The imaging techniques described above all have their disadvantages. These can be partly overcome by the combination of different imaging techniques, for example, by the combination of SPECT with CT.

Although anatomical localisation of radioactivity is improved in SPECT compared with planar lymphoscintigraphy, the absence of anatomical detail on the scans makes it difficult to pinpoint and accurately identify the lymphatics within the body. To solve this problem, in the last decade scanners incorporating both SPECT cameras and conventional CT scanners have been developed [48]. This allows SPECT images to be superimposed on CT images, taken at the same time and with the patient in the same position. Despite the extra radiation dose received by patients from the CT scan, the benefits of being able to pinpoint the exact position of the lymphatics within the body outweigh this disadvantage. Studies on the identification of sentinel lymph nodes have shown the benefit of SPECT/CT over lymphoscintigraphy [49].

The availability of nano-SPECT/CT scanners enables this technique to be used on small animals in the experimental setting, allowing the study of lymphatics in animal models. Our laboratory used SPECT/CT imaging to investigate lymphatic flow in a mouse cardiac transplantation model [50]. Donor hearts were transplanted heterotopically into the abdomen (the usual site for experimental heart transplantation) or subcutaneously into the neck of recipients. One or twenty-eight days after transplantation (timepoints were chosen to be before and after reconnection of the donor lymphatic vessels to those of the recipient as described in the canine model [6]) the hearts were exposed and Tc-99m-labelled human albumin was injected into the myocardial tissue. When transplanted into the abdomen, lymphatic drainage of the donor heart could be seen within the peritoneal cavity 15 min after injection suggesting that the lymph from the donor organ drains freely into the peritoneal cavity. After 6 h, Tc-99m could be detected in the caudal mediastinal lymph nodes, the same route demonstrated from the peritoneal cavity in non-transplantation models (Fig. 7.1). When transplanted into the neck, lymphatic leak could be detected in the surrounding tissue draining into local superficial lymph nodes on the anterior chest wall. SPECT/CT scanning 28 days after transplantation demonstrated that lymphatic leak into surrounding tissue had ceased. The pattern of lymphatic drainage of heart grafts transplanted into the abdomen remained similar, but heart grafts in the neck now drained into deep cervical lymph nodes. The scanning showed that after transplantation, lymph flows freely from the cut ends of the lymphatics, rather than causing blockages in the lymphatics (as a result of the surgical procedure) which could lead to the formation of lymphoceles, as is often seen in patients following renal transplantation. The use of SPECT/CT identified possible sites for the

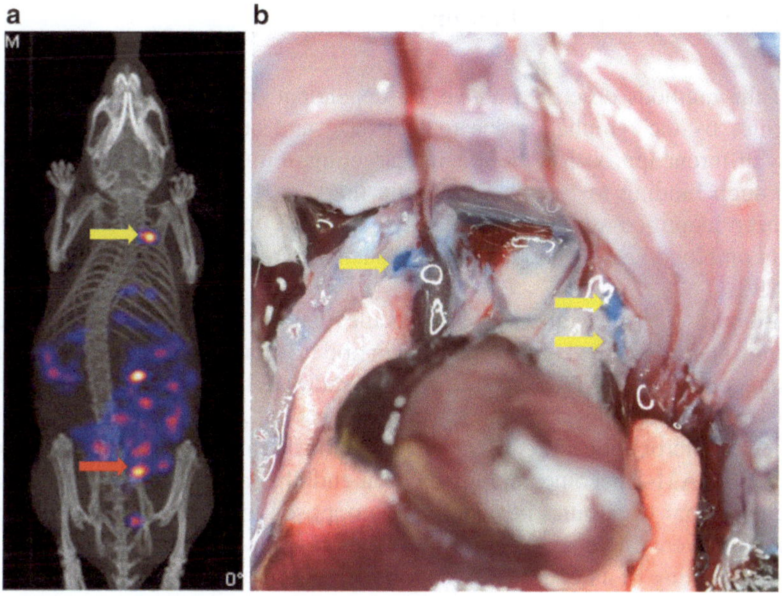

Fig 7.1 (**a**) SPECT/CT images taken 5 h after injection of nanocoll into the wall of a heart transplanted into the abdomen of the recipient. Radioactivity is seen at the site of injection (*red arrow*) and in the mediastinal lymph nodes in the thoracic cavity (*yellow arrow*). (**b**) Injection of Evans blue into the wall of a heart transplanted into the abdomen of the recipient confirms the drainage to the mediastinal lymph nodes (*yellow arrows*).

induction of the immune response that are normally overlooked, and further investigation showed that T cell responses against the graft were initiated in the lymph nodes found through SPECT/CT imaging.

Imaging of lymphatics after cardiac transplantation is unlikely to be performed on its own. By using different imaging techniques, much information can be gained about the function of the heart graft and therefore about whether rejection is taking place and its severity. Many different imaging techniques have been utilised to image heart function. The nuclear imaging techniques SPECT and positron emission tomography (PET) [51], and ultrasound, have all been used to monitor heart function. Both SPECT and ultrasound have shown promise in the diagnosis of rejection in heart transplant patients [52–58]. It is hoped that these non-invasive imaging techniques may replace, at least partially, the need for biopsy of the allograft to monitor rejection. Lymphatic imaging could be combined with the analysis of heart allograft function to give a clearer picture of the alloresponse and predict the likelihood of a rejection episode.

Targeting Lymphatic Markers

All of the methods described above are dependent on the uptake of the imaging molecules by the lymphatics due to their size and other characteristics. A completely different approach is the specific targeting of imaging agents to the lymphatics through binding to lymphatic-specific markers. Many cell surface molecules have been identified as being specifically expressed on lymphatic endothelial cells and not on blood vessel endothelial cells, such as podoplanin, LYVE-1, Prox-1 and VEGFR3.

Antibodies raised against these molecules can be conjugated to many of the imaging agents described above and used in optical imaging (conjugation to fluorescent molecules), SPECT/CT (conjugation to radionuclides) and ultrasound (conjugation to microbubbles). In the case of ultrasound, an anti-ICAM-1 antibody attached to microbubbles was injected intravenously into rat heart transplant recipients [59]. ICAM-1 becomes upregulated on endothelial cells after their activation. A higher signal was found in rejecting, as opposed to non-rejecting, heart grafts. The authors were thus able to demonstrate the principle of using ultrasound to detect antibody binding to an endothelial cell marker. If an antibody against a lymphatic endothelial-specific marker were used, it should be possible to image lymphatic vessels in the same way. The targeting of imaging agents in this way relieves the need for direct injection into the heart. Labelled agents could be injected intravenously with eventual localisation to the lymphatics.

Antibodies have also been widely used to image lymphatics in the traditional way, on heart transplant biopsy sections. The expression of various lymphatic endothelial cell markers during the first year after clinical heart transplantation has been studied [5]. The expression of both LYVE-1 and Prox-1 was found to decrease after transplantation, while VEGFR3 expression did not. A correlation was also found between lower expression of VEGFR3 and more severe rejection. The same group has used similar techniques to study lymphangiogenesis in terminal heart failure, showing the relevance of studying lymphatics in many diseases of the heart, not just post-transplantation [60]. Similar to this, a decrease in the density of LYVE-1 expressing lymphatic vessels has been shown in rejecting heart allografts in a rat transplantation model [7]. The antibodies used in these studies, bound to imaging agents, could be used in the techniques described above.

Data from transplantation of organs other than the heart also provide information on how the lymphatic system responds to transplantation. Lymphangiogenesis has been demonstrated after human renal transplantation, with new lymphatic vessels associated with tertiary lymphoid organs that develop within the graft [9, 10]. Biopsies from kidney transplant patients have also provided information on the source of the new lymphatic vessels, showing that circulating lymphatic progenitor cells from the recipient are incorporated into the new vessels [3]. Blocking lymphangiogenesis following islet transplantation in mice inhibits the alloimmune response and results in islet preservation [12].

Such studies have provided vital information on the involvement of lymphatics in the alloresponse, but it is the need for serial biopsies of the graft that the development of non-invasive imaging techniques is designed to circumvent.

Using Imaging to Asses/Target Therapy

The work described above and that by others [61] has raised the possibility of targeting lymphangiogenesis therapeutically, using, for example, an anti-VEGFR3 antibody to inhibit the development of new lymphatic vessels. This is an intriguing possibility, and imaging of the lymphatic vessels using the techniques described above would allow the effectiveness of this therapy to be monitored and enable the testing of new agents designed to target the lymphatic system. Apart from blue dye injection, all of these techniques allow quantification of lymphatic flow, by serial scanning of the area and measuring either the rate of disappearance of the imaging agent from the site of injection, or the rate of appearance of the imaging agent in the draining lymph node. This aspect of lymphatic imaging would allow the testing of any therapy targeted at lymphangiogenesis as well as the monitoring of its success in individual patients.

As well as specifically targeting lymphangiogenesis to prevent rejection of heart allografts, it may be possible to use the imaging agents described above to deliver current (and future) immunosuppressive therapies into the lymphatic system. Currently, immunosuppression is administered systemically, which leaves the recipient susceptible to infection and cancer. Targeting immunosuppression only to the transplant is an aspect of transplant immunobiology that is being investigated to try and alleviate this problem. The characteristics of the molecules used to image lymphatics, which enable them to be taken up by the lymphatic system draining the tissue they are injected into, unlocks the possibility that these molecules could be conjugated to immunosuppressive drugs to enable these drugs to be concentrated at the site of the alloimmune response, i.e. within the lymphatic system draining the graft. Cyclosporine A (an immunosuppressive drug used in many transplant centres) has been successfully loaded into Tc-99m-labelled dextran acetate particles [62]. When injected subcutaneously into the footpad of rats, the cyclosporine A-loaded particles showed the same distribution pattern as Tc-99m-labelled human serum albumin. This technique could be used not only to deliver drugs to the lymphatic system but also to monitor the distribution and levels of drugs in the body after injection. While it would be difficult to administer repeated doses of drugs, in the same way as it is not ideal to deliver imaging agents into the transplanted heart graft, it would be simple to deliver the first dose of a long acting immunosuppressive agent during surgery to protect the donor graft against the initial wave of immune attack which is usually the most vigorous.

Conclusion

The importance of the immune system following organ transplantation is increasingly being recognised by transplant clinicians and scientists alike. Imaging of the lymphatic system would provide another useful tool with which to study the alloresponse after transplant and in many other cardiac diseases where the lymphatic system is involved. Many techniques have been developed recently to improve imaging of the lymphatic system, and other techniques such as photoacoustic imaging and the use of upconverting nanoparticles for optical imaging are also currently in development. The tools described above, and those which are sure to be available in the near future, should allow the non-invasive monitoring of this important system to become available in the clinic.

Acknowledgements The authors acknowledge the support of the MRC Centre for Transplantation. Work in the authors' laboratory relating to the topics discussed here was funded/supported by the National Institute for Health Research (NIHR) Biomedical Research Centre based at Guy's and St Thomas' NHS Foundation Trust and King's College London. The views expressed are those of the author(s) and not necessarily those of the NHS, the NIHR or the Department of Health. The work was performed in collaboration with Drs Greg Mullen and Kavitha Sunassee in the Division of Imaging Sciences, King's College London. The SPECT/CT scanner used was funded by a grant from the Wellcome Trust.

References

1. Klebe S, Coster DJ, Williams KA (2009) Rejection and acceptance of corneal allografts. Curr Opin Organ Transplant 14(1):4–9
2. Vass DG, Hughes J, Marson LP (2009) Restorative and rejection-associated lymphangiogenesis after renal transplantation: friend or foe? Transplantation 88:1237–1239
3. Kerjaschki D, Huttary N, Raab I, Regele H, Bojarski-Nagy K, Bartel G et al (2006) Lymphatic endothelial progenitor cells contribute to de novo lymphangiogenesis in human renal transplants. Nat Med 12(2):230–234
4. Ochando JC, Homma C, Yang Y, Hidalgo A, Garin A, Tacke F et al (2006) Alloantigen-presenting plasmacytoid dendritic cells mediate tolerance to vascularized grafts. Nat Immunol 7(6):652–662
5. Geissler HJ, Dashkevich A, Fischer UM, Fries JW, Kuhn-Regnier F, Addicks K et al (2006) First year changes of myocardial lymphatic endothelial markers in heart transplant recipients. Eur J Cardiothorac Surg 29(5):767–771
6. Ruggiero R, Fietsam R Jr, Thomas GA, Muz J, Farris RH, Kowal TA et al (1994) Detection of canine allograft lung rejection by pulmonary lymphoscintigraphy. J Thorac Cardiovasc Surg 108(2):253–258
7. Rinda Soong T, Pathak AP, Asano H, Fox-Talbot K, Baldwin WM 3rd (2010) Lymphatic injury and regeneration in cardiac allografts. Transplantation 89(5):500–508
8. Tabibiazar R, Cheung L, Han J, Swanson J, Beilhack A, An A et al (2006) Inflammatory manifestations of experimental lymphatic insufficiency. PLoS Med 3(7):e254
9. Kerjaschki D, Regele HM, Moosberger I, Nagy-Bojarski K, Watschinger B, Soleiman A et al (2004) Lymphatic neoangiogenesis in human kidney transplants is associated with immunologically active lymphocytic infiltrates. J Am Soc Nephrol 15(3):603–612

10. Stuht S, Gwinner W, Franz I, Schwarz A, Jonigk D, Kreipe H et al (2007) Lymphatic neoangiogenesis in human renal allografts: results from sequential protocol biopsies. Am J Transplant 7(2):377–384
11. Brown K, Sacks SH, Wong W (2011) Tertiary lymphoid organs in renal allografts can be associated with donor-specific tolerance rather than rejection. Eur J Immunol 41(1):89–96
12. Yin N, Zhang N, Xu J, Shi Q, Ding Y, Bromberg JS (2011) Targeting lymphangiogenesis after islet transplantation prolongs islet allograft survival. Transplantation 92(1):25–30
13. Santos AC, de Lima JJ, Botelho MF, Pacheco MF, Sousa P, Bernardo J et al (1998) Cardiac lymphatic dynamics after ischemia and reperfusion—experimental model. Nucl Med Biol 25(7):685–688
14. Szlavy L, Koster K, de Courten A, Hollenberg NK (1987) Early disappearance of lymphatics draining ischemic myocardium in the dog. Angiology 38(1 Pt 2):73–84
15. Miller AJ, DeBoer A, Palmer A (1992) The role of the lymphatic system in coronary atherosclerosis. Med Hypotheses 37(1):31–36
16. Lucarelli RT, Ogawa M, Kosaka N, Turkbey B, Kobayashi H, Choyke PL (2009) New approaches to lymphatic imaging. Lymphat Res Biol 7(4):205–214
17. Zhang F, Niu G, Lu G, Chen X (2011) Preclinical lymphatic imaging. Mol Imaging Biol 13(4):599–612
18. Patek PR (1939) The morphology of the lymphatics of the mammalian heart. Am J Anat 64:204–239
19. Feola M, Merklin R, Cho S, Brockman SK (1977) The terminal pathway of the lymphatic system of the human heart. Ann Thorac Surg 24(6):531–536
20. Servelle M, Andrieux J, Cornu C, Deloche A, Nussaume O (1967) [The lymphatics of the heart (peroperative injections)]. Arch Mal Coeur Vaiss 60(1):89–106 [French]
21. Sevick-Muraca EM (2012) Translation of near-infrared fluorescence imaging technologies: emerging clinical applications. Annu Rev Med 63:217–231
22. Proulx ST, Luciani P, Derzsi S, Rinderknecht M, Mumprecht V, Leroux JC et al (2010) Quantitative imaging of lymphatic function with liposomal indocyanine green. Cancer Res 70(18):7053–7062
23. Weiler M, Kassis T, Dixon JB (2012) Sensitivity analysis of near-infrared functional lymphatic imaging. J Biomed Opt 17(6):066019
24. Kim S, Lim YT, Soltesz EG, De Grand AM, Lee J, Nakayama A et al (2004) Near-infrared fluorescent type II quantum dots for sentinel lymph node mapping. Nat Biotechnol 22(1):93–97
25. Kosaka N, Ogawa M, Sato N, Choyke PL, Kobayashi H (2009) In vivo real-time, multicolor, quantum dot lymphatic imaging. J Invest Dermatol 129(12):2818–2822
26. Kobayashi H, Koyama Y, Barrett T, Hama Y, Regino CA, Shin IS et al (2007) Multimodal nanoprobes for radionuclide and five-color near-infrared optical lymphatic imaging. ACS Nano 1(4):258–264
27. Valdés Olmos RA, Hoefnagel CA, Nieweg OE, Jansen L, Rutgers EJ, Borger J et al (1999) Lymphoscintigraphy in oncology: a rediscovered challenge. Eur J Nucl Med 26(4 Suppl):S2–S10
28. Castronuovo JJ Jr, Lopez-Majano V, Flanigan P, Schuler JJ, Jonasson O (1983) Cardiovascular lymphoscintigraphy. Surgery 94(2):351–357
29. Alazraki NP, Eshima D, Eshima LA, Herda SC, Murray DR, Vansant JP, Taylor AT (1997) Lymphoscintigraphy, the sentinel node concept, and the intraoperative gamma probe in melanoma, breast cancer, and other potential cancers. Semin Nucl Med 27(1):55–67
30. Weissleder H, Weissleder R (1988) Lymphedema: evaluation of qualitative and quantitative lymphoscintigraphy in 238 patients. Radiology 167(3):729–735
31. Madeddu G, Spanu A (2004) Use of tomographic nuclear medicine procedures, SPECT and pinhole SPECT, with cationic lipophilic radiotracers for the evaluation of axillary lymph node status in breast cancer patients. Eur J Nucl Med Mol Imaging 31(Suppl 1):S23–S34
32. Spanu A, Tanda F, Dettori G, Manca A, Chessa F, Porcu A et al (2003) The role of (99 m) Tc-tetrofosmin pinhole-SPECT in breast cancer non palpable axillary lymph node metastases detection. Q J Nucl Med 47(2):116–128

33. Herman GT (2009) Fundamentals of computerized tomography: image reconstruction from projections. Springer, London
34. Som PM (1992) Detection of metastasis in cervical lymph nodes: CT and MR criteria and differential diagnosis. AJR Am J Roentgenol 158(5):961–969
35. Suga K, Ogasawara N, Okada M, Matsunaga N (2003) Interstitial CT lymphography-guided localization of breast sentinel lymph node: preliminary results. Surgery 133(2):170–179
36. Johnston DL, Liu P, Wismer GL, Rosen BR, Stark DD, New PF et al (1985) Magnetic resonance imaging: present and future applications. Can Med Assoc J 132(7):765–777
37. Scherzinger AL, Hendee WR (1985) Basic principles of magnetic resonance imaging—an update. West J Med 143(6):782–792
38. Ruehm SG, Corot C, Debatin JF (2001) Interstitial MR lymphography with a conventional extracellular gadolinium-based agent: assessment in rabbits. Radiology 218(3):664–669
39. Weissleder R, Elizondo G, Wittenberg J, Lee AS, Josephson L, Brady TJ (1990) Ultrasmall superparamagnetic iron oxide: an intravenous contrast agent for assessing lymph nodes with MR imaging. Radiology 175(2):494–498
40. Trubetskoy VS, Cannillo JA, Milshtein A, Wolf GL, Torchilin VP (1995) Controlled delivery of Gd-containing liposomes to lymph nodes: surface modification may enhance MRI contrast properties. Magn Reson Imaging 13(1):31–37
41. Koyama Y, Talanov VS, Bernardo M, Hama Y, Regino CA, Brechbiel MW et al (2007) A dendrimer-based nanosized contrast agent dual-labeled for magnetic resonance and optical fluorescence imaging to localize the sentinel lymph node in mice. J Magn Reson Imaging 25(4):866–871
42. Kanno S, Wu YJ, Lee PC, Dodd SJ, Williams M, Griffith BP, Ho C (2001) Macrophage accumulation associated with rat cardiac allograft rejection detected by magnetic resonance imaging with ultrasmall superparamagnetic iron oxide particles. Circulation 104(8):934–938
43. Wu YL, Ye Q, Foley LM, Hitchens TK, Sato K, Williams JB, Ho C (2006) In situ labeling of immune cells with iron oxide particles: an approach to detect organ rejection by cellular MRI. Proc Natl Acad Sci USA 103(6):1852–1857
44. Ye Q, Wu YL, Foley LM, Hitchens TK, Eytan DF, Shirwan H, Ho C (2008) Longitudinal tracking of recipient macrophages in a rat chronic cardiac allograft rejection model with noninvasive magnetic resonance imaging using micrometer-sized paramagnetic iron oxide particles. Circulation 118(2):149–156
45. Christen T, Nahrendorf M, Wildgruber M, Swirski FK, Aikawa E, Waterman P et al (2009) Molecular imaging of innate immune cell function in transplant rejection. Circulation 119(14):1925–1932
46. Flögel U, Su S, Kreideweiss I, Ding Z, Galbarz L, Fu J et al (2011) Noninvasive detection of graft rejection by in vivo (19) F MRI in the early stage. Am J Transplant 11(2):235–244
47. Voigt JU (2009) Ultrasound molecular imaging. Methods 48(2):92–97
48. Bocher M, Balan A, Krausz Y, Shrem Y, Lonn A, Wilk M, Chisin R (2000) Gamma camera-mounted anatomical X-ray tomography: technology, system characteristics and first images. Eur J Nucl Med 27(6):619–627
49. Even-Sapir E, Lerman H, Lievshitz G, Khafif A, Fliss DM, Schwartz A et al (2003) Lymphoscintigraphy for sentinel node mapping using a hybrid SPECT/CT system. J Nucl Med 44(9):1413–1420
50. Brown K, Badar A, Sunassee K, Fernandes MA, Shariff H, Jurcevic S et al (2011) SPECT/CT lymphoscintigraphy of heterotopic cardiac grafts reveals novel sites of lymphatic drainage and T cell priming. Am J Transplant 11(2):225–234
51. Dobrucki LW, Sinusas AJ (2010) PET and SPECT in cardiovascular molecular imaging. Nat Rev Cardiol 7(1):38–47
52. Verhoeven PP, Lee FA, Ramahi TM, Franco KL, Mendes de Leon C, Amatruda J et al (1996) Amatruda J et al. Prognostic value of noninvasive testing one year after orthotopic cardiac transplantation. J Am Coll Cardiol 28(1):183–189

53. Hacker M, Tausig A, Romüller B, Hoyer X, Klauss V, Stempfle U et al (2005) Dobutamine myocardial scintigraphy for the prediction of cardiac events after heart transplantation. Nucl Med Commun 26(7):607–612
54. Ciliberto GR, Ruffini L, Mangiavacchi M, Parolini M, Sara R, Massa D et al (2001) Resting echocardiography and quantitative dipyridamole technetium-99m sestamibi tomography in the identification of cardiac allograft vasculopathy and the prediction of long-term prognosis after heart transplantation. Eur Heart J 22(11):964–971
55. Manrique A, Bernard M, Hitzel A, Bubenheim M, Tron C, Agostini D et al (2010) Diagnostic and prognostic value of myocardial perfusion gated SPECT in orthotopic heart transplant recipients. J Nucl Cardiol 17(2):197–206
56. Wenning C, Stypmann J, Papavassilis P, Sindermann J, Schober O, Hoffmeier A et al (2012) Left ventricular dilation and functional impairment assessed by gated SPECT are indicators of cardiac allograft vasculopathy in heart transplant recipients. J Heart Lung Transplant 31(7):719–728
57. Dandel M, Hummel M, Müller J, Wellnhofer E, Meyer R, Solowjowa N et al (2001) Reliability of tissue Doppler wall motion monitoring after heart transplantation for replacement of invasive routine screenings by optimally timed cardiac biopsies and catheterizations. Circulation 104(12 Suppl 1):I184–I191
58. Bader FM, Islam N, Mehta NA, Worthen N, Ishihara S, Stehlik J et al (2011) Noninvasive diagnosis of cardiac allograft rejection using echocardiography indices of systolic and diastolic function. Transplant Proc 43(10):3877–3881
59. Weller GE, Lu E, Csikari MM, Klibanov AL, Fischer D, Wagner WR, Villanueva FS (2003) Ultrasound imaging of acute cardiac transplant rejection with microbubbles targeted to intercellular adhesion molecule-1. Circulation 108(2):218–224
60. Dashkevich A, Bloch W, Antonyan A, Fries JU, Geissler HJ (2009) Morphological and quantitative changes of the initial myocardial lymphatics in terminal heart failure. Lymphat Res Biol 7(1):21–27
61. Tammela T, Alitalo K (2010) Lymphangiogenesis: Molecular mechanisms and future promise. Cell 140(4):460–476
62. Kim J, Chung KH, Lee CM, Seo YS, Song HC, Lee KY (2008) Lymphatic delivery of 99mTc-labeled dextran acetate particles including cyclosporine A. J Microbiol Biotechnol 18(9):1599–1605

Part III
Cardiac Lymphatic Signaling

Chapter 8
Tie Receptor Signaling in Cardiac Lymphangiogenesis

Xianghu Qu and H. Scott Baldwin

Abstract The Tie receptors and their angiopoietin (Ang) ligands have been identified as the second vascular tissue-specific receptor tyrosine kinase system, following the discovery of the VEGF signaling cascade. In the Ang–Tie system, Ang2 is the first molecule that had been reported to be involved in the regulation of lymphatic development. *Ang2*-deficient mice have profound lymphatic patterning defects. Their lymphatics are characterized by abnormal mural cell recruitment, which perturbs lymphatic function. Interestingly, genetic targeting of Ang1 (an agonistic Tie2 ligand) into the *Ang2* locus restores the lymphatic defects of *Ang2*-deficient mice, which suggests an agonistic mode of action of Ang2 in the lymphatic tissue. The orphan receptor Tie1 is also required for development and function of the lymphatic system. Reduction of Tie1 levels in mice with a hypomorphic *Tie1* allele resulted in abnormal lymphatic patterning, as well as dilated pericardial lymphatics which exhibited disorganized regression. Recent studies in vitro show that Tie1 and Tie2 appear to associate on the cell surface with receptor phosphorylation correlating with activation. Moreover, Tie1 appears to act as an inhibitory co-receptor for Tie2 activation, although the precise mechanisms of Tie receptor signaling in the cardiac lymphatics have yet to be determined.

Keywords Tie1 • Tie2 • Angiopoietins • Lymphatic development • Endothelial cells • Cardiac

---

X. Qu (✉)
Department of Pediatrics (Cardiology), Vanderbilt University Medical Center,
9435-A MRB IV-Langford, 2213 Garland Avenue, Nashville, TN 37232-0493, USA
e-mail: xianghu.qu@vanderbilt.edu

H.S. Baldwin
Division of Cardiology, Vanderbilt Children's Hospital, 9435-A MRB IV-Langford,
2213 Garland Avenue, Nashville, TN 37232-0493, USA
e-mail: scott.baldwin@vanderbilt.edu

G. Karunamuni (ed.), *The Cardiac Lymphatic System: An Overview*,
DOI 10.1007/978-1-4614-6774-8_8, © Springer Science+Business Media New York 2013

Introduction

Despite having been described over 350 years ago, the developmental morphogenesis and subsequent function of the cardiac lymphatic system has received little attention. Current knowledge is summarized in several excellent reviews [1–3]. Recent detailed immunohistochemical studies of lymphatic development in both avian and mouse models suggest that cardiac lymphatic development is a very complex process, which differs from the normal process of lymphangiogenesis characterized by budding of lymphatic endothelial cells from the anterior cardinal vein in the early embryo. There are at least four distinct populations of cells that contribute to the development of the cardiac lymphatics, and two of these cell populations arise external to the developing heart. First, a population of cells coalesces at the arterial pole (aorta and pulmonary trunk) and spreads within the epicardium to cover the heart as a branching lymphatic network. Second, these cells come in contact with isolated cells within the epicardium that are ultimately incorporated in the vascular network. Third, a resident population of lymphatic precursor cells is distributed throughout the myocardium. These cells ultimately differentiate and are recruited into the lymphatic system. Finally, there is a distinct population of cells at the venous pole that invade the heart via the epicardial space in the atrioventricular sulcus and then spread out to cover the dorsal surface of the heart [4–6]. These lymphatic precursors ultimately remodel to form a mature lymphatic circulation comprised of both a lymphatic capillary plexus and collecting lymphatic vessels. A diffuse lymphatic plexus is seen throughout the myocardium, and a plexus also runs between the endocardium and conduction system in the subendocardial space. These fine plexi supply the larger collecting lymphatics which are located in the subepicardial space. The collecting lymphatics coalesce as multiple lymphatic branches which ultimately drain the entire heart via the mediastinal lymphatic vessels including the right lymphatic duct and thoracic duct. Lymphatic valves are prominent in the collecting lymphatic vessels and ensure unidirectional lymph flow. Surprisingly, lymphatic vessels are also found in the atrioventricular and semilunar valves.

While the primary role of lymphatics is to return proteins, lipids, and interstitial tissue fluid to the venous system, the specific function of lymphatics in the heart is not well studied. It is known that acute lymphatic obstruction in dogs results in severe lymphedema, subsequent blistering, and subendocardial hemorrhage with associated myofibrillar degeneration [7]. In addition, acute obstruction results in a depression in both left ventricular systolic and diastolic function [8]. While the exact mechanism of these changes is not understood, several investigators have documented that acute blockade of cardiac lymph results in histological changes [9] as well as ECG changes [10] similar to that seen during ischemia. Others have actually documented evidence of coronary microvascular injury [11]. In keeping with this observation, administration of hyaluronic acid which has been shown to enhance cardiac lymph flow by maintaining lymphatic vascular patency results in decreased myocardial injury following LAD ligation [12] or ischemia reperfusion [13]. Interestingly, chronic cardiac lymph obstruction results not only in depressed myocardial function of the right and left ventricles but also leads to myxoid changes along with fibrosis in the atrioventricular valves. This phenomenon is particularly prominent in the tricuspid valves [14, 15].

The critical role of the VEGFR-3 receptor and its ligands VEGF-C and VEGF-D, as well as several other signaling pathways involving Sox18, Prox1, Coup-TFII, neuropilin-2, ephrin B2, and Notch in lymphatic development and vascular disease, have been recently reviewed [16, 17]. However, few of these studies have delineated the role of these signaling pathways in the complicated process of cardiac lymphatic development. We have recently described a critical role for the Tie family of receptor tyrosine kinases (RTKs) in cardiac lymphatic development, and this review will focus on this unique family of receptors and ligands.

Structure and Expression of Tie Receptors and Angiopoietin Ligands

Following the discovery of the VEGF–VEGFR system, the Tie (tyrosine kinase with immunoglobulin-like and EGF homology) receptors, together with their corresponding angiopoietin (Ang) ligands, were identified as the second endothelial cell (EC)-specific receptor Tyr kinase signaling system [18–20]. The Tie receptors (Tie1 and Tie2 (also known as TEK)) were originally identified as orphan receptors. Ang1 (also known as Angpt1) and Ang2 (also known as Angpt2) were subsequently cloned as agonistic and antagonistic Tie2 ligands, respectively [21, 22] acting in either a paracrine (Ang1) or autocrine (Ang2) manner. Loss-of-function and gain-of-function experiments of Tie receptors and Ang ligands have established the Ang–Tie system as a gatekeeper of the quiescent EC phenotype. As such, the system plays a key role in remodeling and maturation of blood vessels and is particularly important for lymphatic development [23]. The Tie signaling system has many important parallels to the better understood VEGF system. For example, similar to what has been found with VEGF receptors, the Tie receptors are expressed selectively by endothelial cells. Signaling by Tie receptors appears to complement the VEGF pathway by contributing to later stages of vascular development. Thus, whereas VEGF signals promote initiating events in angiogenesis such as EC sprouting, Ang–Tie signals appear to promote EC survival and vascular assembly, stability, and maturation.

Correspondingly, the Tie signaling system is involved in pathological processes that involve the vessel wall, ranging from short-term adaptive processes, such as inflammation, to more protracted long-term vascular functions, such as tumor angiogenesis. For example, heritable venous malformations in two families were found to be associated with a missense mutation in the kinase domain of Tie2 [24]. This mutation results in increased activity of Tie2. Subsequent studies have found that sporadic venous malformations are also associated with somatic point mutations in Tie2, which again are activating [25]. Increased understanding of the Ang–Tie system and relative technological developments in lymphatic vascular biology will lead to a better understanding and treatment of these diseases. This chapter provides an update on recent advances in Ang–Tie biology and focuses on our current understanding of the Ang–Tie system in the cardiac lymphatics.

Tie Receptors

Tie1 and Tie2 are EC-specific receptors with similar molecular weights of approximately 135 kDa and 150 kDa, respectively. Both receptors are structurally similar in the cytoplasmic region (76 % sequence identity), but show only 33 % similarity in the extracellular part [20]. Each consists of an extracellular domain, transmembrane helix, and intracellular tyrosine kinase domain with kinase insert [18, 26]. The extracellular region of the Tie receptors consists of one complete and one incomplete immunoglobulin (lg)-like motifs at the N-terminus that are separated by three epidermal growth factor (EGF) homology domains and followed by three fibronectin (FN) III-like repeats before the transmembrane domain (Fig. 8.1). Crystal structure analyses reveal that the ligands Ang1 and Ang2 bind with almost similar affinity to the same site (the second Ig motif) of the Tie2 receptor [27, 28]. Despite the similarities in ectodomain structure between the two receptors, Tie1 is unable to bind native angiopoietins [21, 22]. However, Tie1 interacts with Tie2 on the surface of ECs and modulates Tie2 activation by angiopoietin ligands [29–31].

Tie2 is the first of the RTKs to be expressed in the embryo, being detected as early as E8 with Tie1 expression being detected slightly later at E8.5. Expression patterns for these two RTKs are virtually indistinguishable after E8.5 [20, 32]. Tie1 and Tie2 are co-expressed in all blood vascular and lymphatic ECs. In fact, to date no EC population has been described to only express Tie1 or Tie2. Tie receptors are also expressed by circulating hematopoietic cells, including megakaryocytes, as

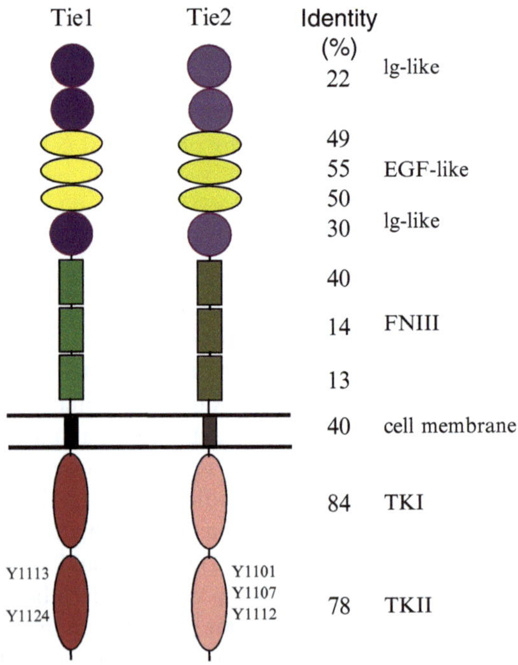

Fig. 8.1 Schematic overview of the Tie receptors. Amino acid identities between mouse Tie1 and Tie2 within each domain are indicated by percentages. Tie2 has three phosphotyrosine residues (1101, 1107, 1112), whereas Tie1 has only two (1113 and 1124). The equivalent to Tie2 pTyr1107 is missing. *Ig* immunoglobulin; *EGF* epidermal growth factor; *FN* fibrinogen; *TK*, tyrosine kinase

well as by hematopoietic stem cells in the bone marrow niche. EC Tie1 mRNA expression is upregulated at sites of turbulent blood flow, and this accentuation of Tie1 expression is associated with the development of atherosclerosis [33, 34].

Angiopoietins

The angiopoietins are a family of secreted glycoproteins with a dimeric molecular weight of approximately 75 kDa. These ligands bind the endothelial RTK Tie2, and this is the primary dedicated receptor for the angiopoietins mediating their effects [21]. There are four angiopoietins known: angiopoietin-1 (Ang1), angiopoietin-2 (Ang2), angiopoietin-3 (Ang3), and angiopoietin-4 (Ang4). Ang1 and Ang2 are the best characterized of these ligands. Ang3 and Ang4 are orthologs found in mouse and human, respectively. Structurally, the angiopoietins are composed of two domains: an N-terminal coiled-coil domain and an N-terminal fibrinogen-like domain which are responsible for ligand homo-oligomerization of the ligands and for receptor activation, respectively [21, 35]. At the amino acid level, Ang1 and Ang2 show sequence homology of about 60 % overall [21, 22].

In contrast to the wide coexpression of Tie receptors by ECs, the Ang ligands have distinct expression patterns. Ang1 is expressed by smooth muscle cells and other perivascular cells and acts in a paracrine manner on the endothelium. It is abundantly expressed by the myocardium during early development and by perivascular cells later during development and in adult tissues [22, 36, 37]. In contrast to Ang1, Ang2 is almost exclusively expressed by ECs. Its expression is transcriptionally induced by factors, such as hypoxia, shear stress, and VEGF, and downregulated by the transcription factor Kruppel-like factor 2 [38–41]. In most normal adult tissues, Ang2 is expressed at low levels but is strongly upregulated at times of both vessel growth and regression and at sites of active vessel remodeling, such as in the ovary [42] and during tumor angiogenesis [43, 44], suggesting that it plays an active role in vessel remodeling.

Physiological Roles of Tie Receptors and Angiopoietin Ligands in the Cardiovascular System

Tie Receptors

Direct evidence from gene-targeted mice and overexpression studies with transgenic mice suggest that the Ang–Tie system plays a key role during vessel remodeling, maturation, and stabilization of the cardiovascular system. However, null mutant embryos of each RTK have dramatically different phenotypes. *Tie2* deficiency resulted in embryonic lethality at approximately embryonic day 10.5 (E10.5), associated with cardiac failure, hemorrhage, and other vascular defects. The heart and major blood vessels are able to form, but the heart fails to undergo normal vascularization and trabeculation [45, 46]. The hearts of Tie2 null mice begin the process of trabeculation but have much reduced interdigitation of ECs and cardiac cells, as well as poor association of the ECs with the myocardium. The initial phases of angiogenesis and blood vessel formation, including sprouting, are able to proceed normally. However, blood vessels throughout the embryo apparently do not remodel or form normal hierarchical networks. In addition, the aortas of E9.5 *Tie2* null mice have fewer ECs [45] and a profound lack of smooth muscle cells; those smooth muscle cells that are present appear to be tenuously attached to the blood vessels. Tie2 also exerts critical roles during hematopoiesis [47].

By contrast, mice deficient for *Tie1* die between E13.5 and E18.5, depending on the background strain [48]. Vascular development proceeds normally in *Tie1*-deficient mice up to approximately E13.0. Shortly thereafter, *Tie1* mutant mice begin to show signs of edema, local hemorrhage, and rupturing of microvessels. However, the major blood vessels appear intact [46, 48]. Thus, Tie1 is not required for vasculogenesis, but it is required for the integrity [49] and survival [50] of vascular ECs. Unlike *Tie2*-deficient mice, hematopoiesis occurs normally in *Tie1*-deficient mice [51].

Interestingly, the lymphatic vasculature appears to be particularly sensitive to the level of Tie1 expression as a hypomorphic allele with reduced expression resulted in normal formation of the lymphatic endothelium but failure of the lymphatic vasculature to remodel following initial formation of the vascular plexus [52]. While lymphatic defects were obvious in the hypomorphic mutant embryos, the pattern of blood vessels often appeared relatively normal.

Angiopoietin Ligands

Ang1 deficiency results in lethality at E11–E12.5 with defects in heart and vascular development [36]. The phenotype is similar to but slightly less severe than those seen in mice lacking *Tie2*. These embryos have growth-retarded hearts with a less complex ventricular endocardium. The endothelial lining in the atria is collapsed and the trabeculae are absent. In addition, *Ang1*-deficient mice exhibit severe vascular defects: a much simpler and immature primary capillary plexus is present, while periendothelial cells are scarce and appear separated from rounded ECs.

Myocardial overexpression of Ang1 under the control of the tetracycline promoter shed further light on the importance of Ang1 during heart development. 90 % of these mice die between E12.5 and E15.5 as a result of cardiac hemorrhage. The transgenic mice exhibit dilated atria, a significant thinning of the myocardial wall, and eventual outflow tract collapse. In addition, hearts of the most severely affected transgenic embryos have no coronary arteries as a result of the defective development and maintenance of the epicardium [53]. Overexpression of Ang1 in the skin using the keratin 14 gene (Krt14) promoter leads to the formation of larger, more numerous, and more highly branched vessels, thus resulting in dermal hypervascularization [54].

Genetic manipulation of *Ang2* in mice has validated its role in vascular development but has also revealed its complexity. The initial studies with genetic manipulation described transgenic overexpression of *Ang2* in ECs (via a *Tie2* promoter). Transgenic overexpression of Ang2 disrupts blood vessel formation in the mouse embryo, which leads to mid-gestational embryonic lethality at ~E9.5–E10 that is similar to the phenotype of *Ang1*- and *Tie2*-deficient mice [22, 36]. The endocardial lining is collapsed and detached from the underlying myocardium. Trabecular folds are completely absent. Furthermore, endothelial-specific overexpression of *Ang2* in adult mice exhibits a complete suppression of Ang1-mediated Tie2 phosphorylation in addition to arteriogenesis defects [55]. These results support the hypothesis that Ang1 acts in a stimulating, agonistic manner on Tie2, whereas Ang2 exerts antagonistic functions on Ang1/Tie2 signaling.

Mice genetically deficient for *Ang2* showed complex lymphatic and vascular phenotypes [37], and the phenotypes are strain dependent. All *Ang2*-deficient mice in the 129/J background die in the first 14 days after birth as a consequence of chylous ascites, while *Ang2*-deficient mice in the C57/Bl6 background develop transient postnatal chylous ascites with only 10 % postnatal lethality [56]. Further analyses revealed that *Ang2*-deficient mice also show widespread but minor vascular dysmorphogenesis in the body [57]. The results from the *Ang2*-deficient mice again indicate a role for angiopoietins in the remodeling phases of both the blood and lymphatic vessels.

Signaling Through Ang–Tie System in Lymphatics

Tie1 and *Tie2* are expressed in lymphatic tissue, although at lower levels compared to blood vessels [37, 58–60] and conversely *Ang2* is expressed more abundantly in lymphatic ECs than in blood ECs. In the Ang–Tie system, Ang2 is the first molecule that had been reported to be involved in the regulation of lymphatic development based on the phenotype of *Ang2*-deficient mice. *Ang2* mutant mice exhibit minor defects in the remodeling of the blood vasculature, but, surprisingly, even more severe defects of the lymphatic system. *Ang2* mutant mice display chylous ascites, peripheral lymphedema, and hypoplasia of the lymphatic vasculature [37]. Lymphatic vessels in *Ang2* mutant mice fail to mature and do not exhibit a collecting vessel phenotype. Furthermore, dermal lymphatic vessels in *Ang2* mutant pups prematurely recruit smooth muscle cells and do not undergo proper postnatal remodeling [61]. Therefore, Ang2 is involved in the remodeling and stabilization of lymphatic vessels.

Strikingly, the genetic knock-in of *Ang1* into the *Ang2* locus completely rescues the lymphatic phenotype of *Ang2*-deficient mice, but not the vascular remodeling defects [37]. Consistent with the redundant roles of Ang1 and Ang2 in the lymphatic system, adenoviral overexpression or recombinant delivery of cartilage oligomeric matrix protein (COMP)-Ang1 stimulates lymphangiogenesis in cellular and in vivo models [59, 60]. These data support the hypothesis that Ang2 is agonistic in lymphatic vessels and antagonistic in blood vessels. It also suggests that Ang2 is dispensable for early development but necessary for vessel remodeling during later stages of development.

Tie1 is expressed in the Prox1-positive venous lymphatic EC progenitors that form the first lymph sacs, and this expression of *Tie1* in the lymphatic system is maintained throughout whole embryonic development to persist in postnatal mice [52]. Recently, two individual reports have demonstrated that Tie1 is also required for development and function of the lymphatic system [52, 62]. Careful analyses of the mice with a hypomorphic Tie1 allele reveal that Tie1 is involved in the remodeling and stabilization of lymphatic vessels [52]. In these mice, the expression level of Tie1 is reduced. The phenotype of these mice reveals a variable severity that is usually milder than the conventional complete knockout mice. Before E16.5, the majority (91.7 %) of the hypomorphic homozygous mutant embryos exhibit mild to severe edema without signs of hemorrhage, while the others appear normal. Rare hypomorphic homozygous mutant embryos die before E18.5 and a few even survive to adulthood, but most mutant mice die perinatally. The Tie1 reduction embryos displayed abnormal lymphatic patterning and dilated and disorganized lymphatic vessels in all tissues examined, including the heart, diaphragm, gut, and lung. Consistent with structural defects seen in lymphatic vessels, lymphatic drainage function is impaired in mutant embryonic skin. The homozygous hypomorphic mutant mice also exhibited abnormal jugular lymphatic vessels associated with an increase in lymphatic EC apoptosis. The severity of the phenotypes observed correlated with the expression levels of Tie1 confirming that there is a dosage dependence for Tie1 in the integrity and survival of developing lymphatic endothelia. In addition, while lymphatic defects were obvious in the hypomorphic mutant embryos, the pattern of blood vessels often appeared relatively normal. These results indicate that the development of lymphatic vasculature is more sensitive to reduced Tie1 levels than the development of blood vasculature.

The *Tie1* hypomorphic mutant embryos with severe edema and hemorrhage defects also had smaller hearts and smaller lungs than wild-type controls, similar to what was seen in the complete knockout mice [46]. Whole-mount staining of these hearts with Lyve-1 or VEGFR3 antibodies revealed regression of the developing lymphatic vessels at the surface of the heart in the mutant embryos compared with control littermates (Figs. 8.2, 8.3, and 8.4). Normally lymphatic capillaries progressively cover the heart surface and the diaphragm from E15 onwards [63], while the phenotype in *Tie1* hypomorphic mutant embryos was first observed at E16.5. The epicardial lymphatic vessels in the hypomorphic mutant embryos still formed a continuous network but were thinner than those in wild-type control embryos (Fig. 8.2). In addition, the lymphatic vessels of the mutant embryos had begun to regress in some areas. At E17.5, the whole lymphatic network was disrupted in the mutant embryos (Fig. 8.3). By E18.5, no or only a few single disrupted lymphatic vessels were detected at the surface of the heart in the mutant embryos (Fig. 8.4). The abnormal lymphatic patterning of mutant mice is associated with a selective increase in lymphatic EC apoptosis, which is consistent with the observed regression of the already-formed developing lymphatics. These findings also support previous reports that Tie1 is required in a cell autonomous mechanism for endothelial cell survival and that Tie1 inhibits apoptosis [64].

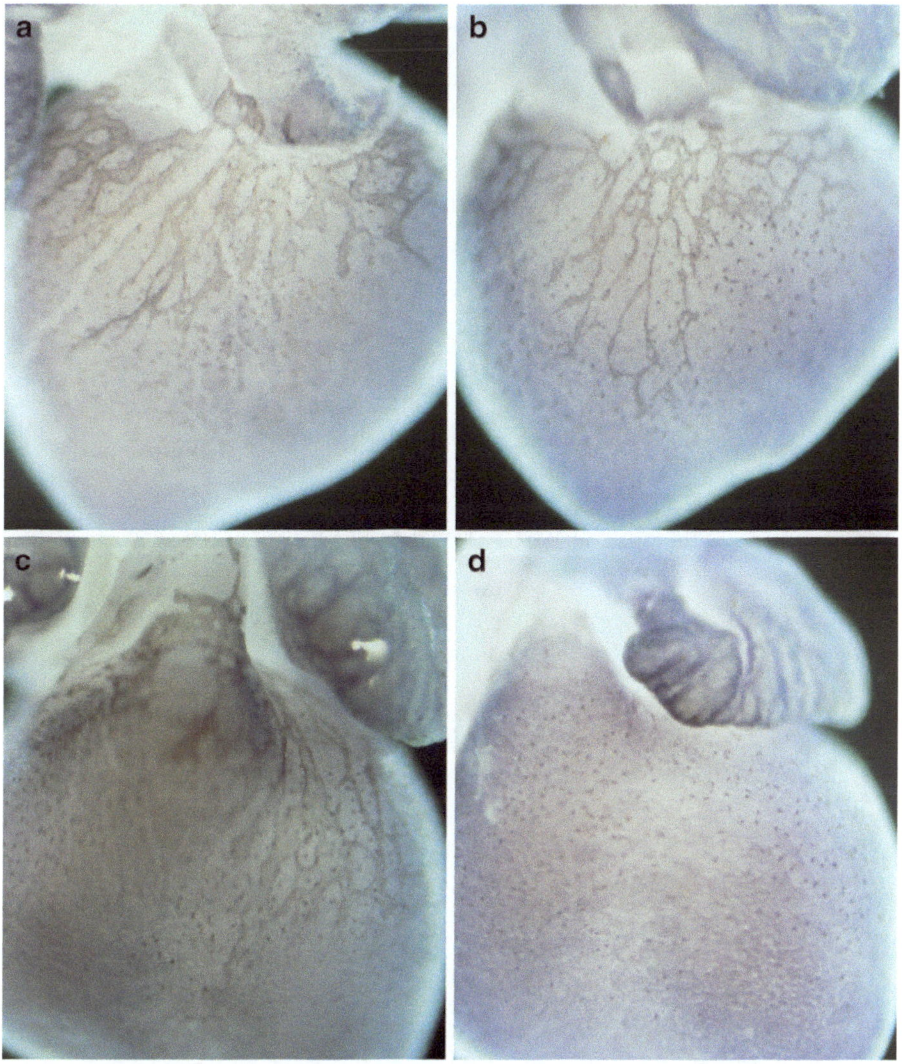

Fig. 8.2 Hearts of WT control (**a**, **c**) and *Tie1* hypomorphic (**b**, **d**) embryos at E16.5 were whole-mount labeled with VEGFR3 antibody. Compared to the normal pericardial lymphatic network in control littermate hearts, the hypomorphic pericardial lymphatic vessels were thinner and had started to regress in some areas. (**a**, **b**) Anterior view and (**c**, **d**) posterior view

Thus, continuous Tie1 signaling is required for the survival of the lymphatic ECs. However, the exact cellular and molecular mechanisms underlying this role remain to be determined. In addition, the mechanism for Tie2 signaling in the lymphatic system needs to be further delineated.

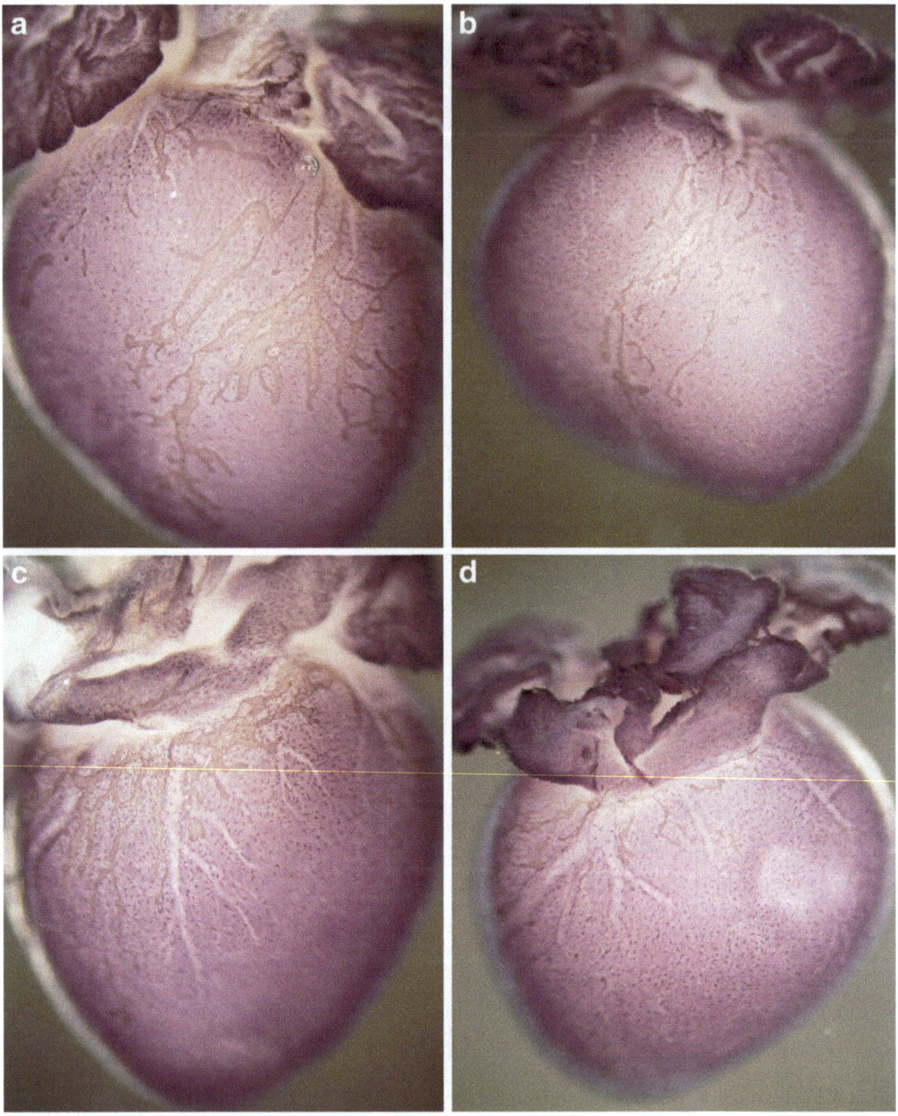

Fig. 8.3 Hearts of WT control (**a**, **c**) and *Tie1* hypomorphic (**b**, **d**) embryos at E17.5 were whole-mount labeled with VEGFR3 antibody. Compared to the normal pericardial lymphatic network in control littermate hearts, the whole lymphatic network was disrupted in the hypomorphic embryos. (**a**, **b**) Anterior view and (**c**, **d**) posterior view

Molecular Regulation of Ang–Tie Signaling

The lymphatic defects of *Tie1* mutant mice are reminiscent of the lymphatic phenotypes of the Tie2 ligand *Ang2* knockout mice. As *Tie1* is expressed in the endothelium of the same structures as *Ang2* during embryogenesis and enhanced *Tie1*

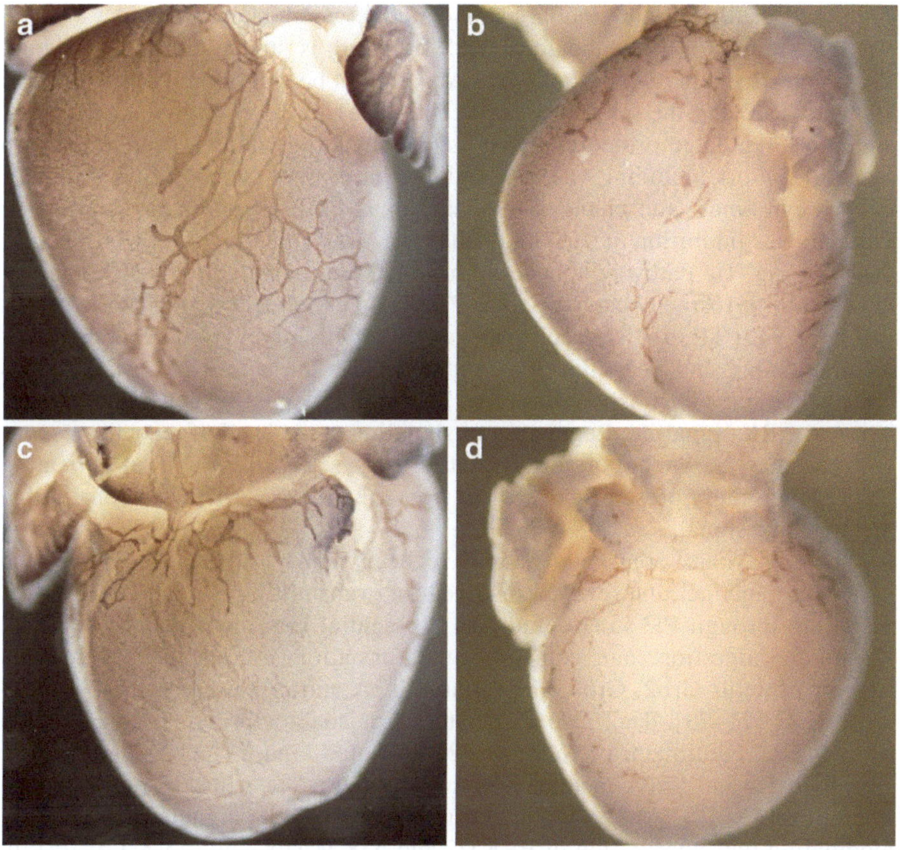

Fig. 8.4 Hearts of WT control (**a**, **c**) and *Tie1* hypomorphic (**b**, **d**) embryos at E18.5 were whole-mount labeled with VEGFR3 antibody. Compared to the normal pericardial lymphatic network in control littermate hearts, only a few single disrupted lymphatic vessels were at the pericardial surface in the hypomorphic embryos. (**a**, **b**) Anterior view, (**c**, **d**) posterior view

mRNA expression occurs together with *Ang2* mRNA in the adults, it is tempting to speculate that Tie1 might be the molecule modulating Ang2 function via the Tie2 receptor in the lymphatic ECs. In fact, increasing evidence support that Tie1 plays a role as an inhibitory co-receptor for Tie2 since Tie1 and Tie2 appear to associate on the cell surface and Tie1 also attenuates Ang1 activation or Ang 2 inhibition of Tie2 activation [31]. In vitro studies suggest that this is in fact the case in mature lymphatic endothelial cells [65]. The absolute requirement for Tie1 and Tie2 during angiogenic remodeling and vessel maintenance suggests that a well-balanced Tie2/Tie1 ratio is critical to the development of new vessels and the regression of existing vessels.

Ang1- or *Tie2*-overexpression experiments have suggested that excess Tie2 signaling might have deleterious effects, as does the loss of Tie2 function. Similarly, a mutation that activates Tie2 causes venous malformations that are composed of dilated endothelial channels [24], which shows that like VEGF–VEGFR signaling, the outcome of Tie2 signaling is strictly dosage regulated. The responsiveness of endothelium to Ang1 might be determined by the relative levels of Tie2 and the inhibitory co-receptor Tie1 in the endothelial cells. Tie1 undergoes regulated proteolytic ectodomain shedding which is stimulated by a range of factors including vascular endothelial growth factor (VEGF), inflammatory cytokines, and changes in shear stress [66]. Ectodomain cleavage of Tie1 relieves inhibition of Tie2 and enhances Ang1 signaling. It is worth noting that the Ang–Tie pathway appears to have another negative regulator, namely, the receptor tyrosine phosphatase VE-PTP (also called PTPRβ). Genetic inactivation of VE-PTP in mice results in embryonic lethality, with defects in vascular remodeling and heart development [67, 68]. The activity of VE-PTP has been linked to its dephosphorylation of Tie2 [69, 70] and subsequent downregulation of ERK signaling. Regulation of signaling at the level of receptor responsiveness may be an important adaptation in systems in which an activating ligand is normally present in excess or where the ligand provides a constitutive maintenance signal.

In recent years, some advances have been made towards elucidating the molecular and cellular events involved in transmitting signals from Tie2. Signaling through PI3 kinase and Akt is essential for Ang1-induced survival, sprouting, migration, and capillary tube formation. Activated Tie2 recruits adaptor proteins Grb2, Grb7, Grb14, and ShcA and protein tyrosine phosphatase SHP2 [71–73]. Tie2 also associates with adaptor protein Dok-R, which leads to the recruitment of Nck and p21-activated kinase (PAK) [74, 75]. The Dok-R-PAK- and ShcA-mediated pathways are involved in the Ang1-induced migration [73, 74]. Ang1 may also regulate the MAPK signaling cascade by modulating phosphorylation of ERK1/2 and p38 MAPKs by PI3-K [76, 77]. In contrast to Tie2 signal transduction, Tie1 signaling is still poorly understood. There is data to support that Tie1 exhibits no measurable tyrosine kinase activity [29]. However, during the course of Tie2 activation, others have observed weak Tie1 phosphorylation [78, 79], although the physiological significance of Tie1 phosphorylation remains unknown. In addition, Tie1–Tie2 interactions might be required for Tie1 activation, and Tie1 could modulate Tie2 signaling by the ectodomain and/or the cytoplasmic domains [30, 31].

It is important to note that Tie signaling is context dependent, and specific mechanisms of activation are dependent on the endothelial subpopulation studied [80]. The models describing the signal-transduction mechanisms of the Tie receptors are based primarily on in vitro studies. Because the lymphatic vascular system is a complex organ system and angiogenesis involves interactions between ECs and numerous other cell types, it will now be important to reevaluate the findings outlined here using innovative approaches to study signaling pathways in vivo. The generation of mice with mutant receptors that block binding to key downstream binding partners might be one way in which researchers can more accurately define the molecular functions of the Tie receptors in the context of a complex cellular environment.

Summary

The development and maintenance of the cardiac lymphatic system is clearly a complex process that involves multiple endothelial populations that are ultimately orchestrated into a functional vascular system required for normal cardiac function. The Tie1-angiopoietin system plays a central role in this process, but more work needs to be done to delineate the specific role of this signaling system as well as other signaling pathways in this critical process. Further delineation of the essential components of lymphatic development and repair during disease, ischemia, and mechanical disruption imposed by surgical interventions may provide avenues for therapeutic intervention to facilitate recovery of myocardial function.

References

1. Cui Y (2010) Impact of lymphatic vessels on the heart. Thorac Cardiovasc Surg 58:1–7
2. Cui Y (2010) The role of lymphatic vessels in the heart. Pathophysiology 17:307–314
3. Loukas M, Abel N, Tubbs RS, Grabska JB, Anderson RH (2011) The cardiac lymphatic system. Clin Anat 24:684–691
4. Juszynski M, Ciszek B, Stachurska E, Jablonska A, Ratajska A (2008) Development of lymphatic vessels in mouse embryonic and early postnatal hearts. Dev Dyn 237:29073–29086
5. Karunamuni G, Yang K, Doughman YQ, Wikenheiser J, Bader D, Barnett J et al (2010) Expression of lymphatic markers during avian and mouse cardiogenesis. Anat Rec 293:259–270
6. Flaht A, Jankowska-Steifer E, Radomska DM, Madej M, Gula MG, Kujawa M et al (2012) Cellular phenotypes and spatio-temporal patterns of lymphatic vessel development in embryonic mouse hearts. Dev Dyn 241:1473–1486
7. Sun S, Lie J (1977) Cardiac lymphatic obstruction. Mayo Clin Proc 52:785–792
8. Ludwig LL, Schertel ER, Pratt JW, McLure DE, Ying AJ, Heck CF et al (1997) Impairment of left ventricular function by acute cardiac lymphatic obstruction. Cardiovasc Res 33:1645–1671
9. Pick A, Miller AJ, Gliock G (1974) Myocardial pathology after cardiac venous and lymph flow obstruction in the dog. Am Heart J 87:627–632
10. Gloviczki P, Solti F, Szlavy L, Jellinek H (1983) Utrastructural and electrophysiologic changes of experimental acute lymphostasis. Lymphology 16:185–192
11. Solti F, Lengyel E, Jellinek H, Schneider F, Jushasz-Nagy A, Kekesi V (1994) Coronary arteriopathy after lymphatic blockade: an experimental study in dogs. Lymphology 27:173–180
12. Taira A, Uehara K, Fukuda S, Takenaka K, Koga M (1990) Active drainage of cardiac lymph in relation to reduction in size of myocardial infarction: an experimental study. Angiology 41:1029–1036
13. Yotsmoto G, Moriyama Y, Yamaoka A, Taira A (1998) Experimental study of cardiac lymph dynamics and edema formation in ischemia/reperfusion injury: with reference to the effect of hyaluronidase. Angiology 49:299–305
14. Symbass PN, Cooper T, Gantner GE Jr, Willman VL (1969) Lymphatics of the heart: anatomic effects following interruption of the drainage of the cardiac lymph. J Thorac Cardiovasc Surg 81:573–575
15. Symbass PN, Schlant RC, Gravanis MB, Shepherd RL (1969) Pathologic and functional effects on the heart following interruption of the cardiac lymph drainage. J Thorac Cardiovasc Surg 57:577–584

16. Alitalo K (2011) The lymphatic vasculature in disease. Nat Med 17:1371–1380
17. Oliver G, Srinivasan RS (2010) Endothelial plasticity: how to become and remain a lymphatic endothelial cell. Development 137:363–372
18. Dumont DJ, Yamaguchi TP, Conlon RA, Rossant J, Breitman ML, Dumont DJ, Yamaguchi TP, Conlon RA, Rossant J, Breitman ML (1992) tek, a novel tyrosine kinase gene located on mouse chromosome 4, is expressed in endothelial cells and their presumptive precursors. Oncogene 7:1471–1480
19. Partanen J, Armstrong E, Mäkelä TP, Korhonen J, Sandberg M, Renkonen R et al (1992) A novel endothelial cell surface receptor tyrosine kinase with extracellular epidermal growth factor homology domains. Mol Cell Biol 12:1698–1707
20. Sato TN, Qin Y, Kozak CA, Audus KL (1993) Tie-1 and tie-2 define another class of putative receptor tyrosine kinase genes expressed in early embryonic vascular system. Proc Natl Acad Sci USA 90(20):9355–9358
21. Davis S, Aldrich TH, Jones PF, Acheson A, Compton DL, Jain V et al (1996) Isolation of angiopoietin-1, a ligand for the TIE2 receptor, by secretion-trap expression cloning. Cell 87:1161–1169
22. Maisonpierre PC, Suri C, Jones PF, Bartunkova S, Wiegand SJ, Radziejewski C et al (1997) Angiopoietin-2, a natural antagonist for Tie2 that disrupts in vivo angiogenesis. Science 277:55–60
23. Wu X, Liu N (2010) The role of Ang/Tie signaling in lymphangiogenesis. Lymphology 42:59–72
24. Vikkula M, Boon LM, Carraway KL III, Calvert JT, Diamonti AJ, Goumnerov B et al (1996) Vascular dysmorphogenesis caused by an activating mutation in the receptor tyrosine kinase TIE2. Cell 87(7):1181–1190
25. Limaye N, Wouters V, Uebelhoer M, Tuominen M, Wirkkala R, Mulliken JB et al (2009) Somatic mutations in angiopoietin receptor gene TEK cause solitary and multiple sporadic venous malformations. Nat Genet 41:118–124
26. Runting AS, Stacker SA, Wilks AF (1993) tie2, a putative protein tyrosine kinase from a new class of cell surface receptor. Growth Factors 9(2):99–105
27. Barton WA, Tzvetkova-Robev D, Miranda EP, Kolev MV, Rajashankar KR, Himanen JP et al (2006) Crystal structures of the Tie2 receptor ectodomain and the angiopoietin-2–Tie2 complex. Nat Struct Mol Biol 13:524–532
28. Fiedler U, Krissl T, Koidl S, Weiss C, Koblizek T, Deutsch U et al (2003) Angiopoietin-1 and angiopoietin-2 share the same binding domains in the Tie-2 receptor involving the first Ig-like loop and the epidermal growth factor-like repeats. J Biol Chem 278:1721–1727
29. Marron MB, Hughes DP, Edge MD, Forder CL, Brindle NPJ (2000) Evidence for heterotypic interaction between the receptor tyrosine kinases Tie1 and Tie2. J Biol Chem 275:39741–39746
30. Marron MB, Singh H, Tahir TA, Kavumkal J, Kim H-Z, Koh GY et al (2007) Regulated proteolytic processing of Tie1 modulates ligand responsiveness of the receptor tyrosine kinase Tie2. J Biol Chem 282:30509–30517
31. Seegar TC, Eller B, Tzvetkova-Robev D, Kolev MV, Henderson SC, Nikolov DB et al (2010) Tie1–Tie2 interactions mediate functional differences between angiopoietin ligands. Mol Cell 37:643–655
32. Dumont DJ, Fong GH, Puri MC, Gradwohl G, Alitalo K, Breitman ML (1995) Vascularization of the mouse embryo: a study of flk-1, tek, tie, and vascular endothelial growth factor expression during development. Dev Dyn 203(1):80–92
33. Porat RM, Grunewald M, Globerman A, Itin A, Barshtein G, Alhonen L et al (2004) Specific induction of tie1 promoter by disturbed flow in atherosclerosis-prone vascular niches and flow-obstructing pathologies. Circ Res 94:394–401
34. Woo KV, Qu X, Babaev VR, Linton MF, Guzman RJ, Fazio S et al (2011) Tie1 attenuation reduces murine atherosclerosis in a dose-dependent and shear stress-specific manner. J Clin Invest 121(4):1624–1635

35. Procopio WN, Pelavin PI, Lee WM, Yeilding NM (1999) Angiopoietin-1 and -2 coiled coil domains mediate distinct homo-oligomerization patterns, but fibrinogen-like domains mediate ligand activity. J Biol Chem 274:30196–30201
36. Suri C, Jones PF, Patan S, Bartunkova S, Maisonpierre PC, Davis S et al (1996) Requisite role of angiopoietin-1, a ligand for the TIE2 receptor, during embryonic angiogenesis. Cell 87:1171–1180
37. Gale NW, Thurston G, Hackett SF, Renard R, Wang Q, McClain J et al (2002) Angiopoietin-2 is required for postnatal angiogenesis and lymphatic patterning, and only the latter role is rescued by Angiopoietin-1. Dev Cell 3:411–423
38. Mandriota SJ, Pepper MS (1998) Regulation of angiopoietin-2 mRNA levels in bovine microvascular endothelial cells by cytokines and hypoxia. Circ Res 83:852–859
39. Mandriota SJ, Pyke C, Di Sanza C, Quinodoz P, Pittet B, Pepper MS (2000) Hypoxia-inducible angiopoietin-2 expression is mimicked by iodonium compounds and occurs in the rat brain and skin in response to systemic hypoxia and tissue ischemia. Am J Pathol 156:2077–2089
40. Oh H, Takagi H, Suzuma K, Otani A, Matsumura M, Honda Y (1999) Hypoxia and vascular endothelial growth factor selectively up-regulate angiopoietin-2 in bovine microvascular endothelial cells. J Biol Chem 274:15732–15739
41. Pichiule P, Chavez JC, LaManna JC (2004) Hypoxic regulation of angiopoietin-2 expression in endothelial cells. J Biol Chem 279:12171–12180
42. Goede V, Schmidt T, Kimmina S, Kozian D, Augustin HG (1998) Analysis of blood vessel maturation processes during cyclic ovarian angiogenesis. Lab Invest 78:1385–1394
43. Holash J, Maisonpierre PC, Compton D, Boland P, Alexander CR, Zagzag D et al (1999) Vessel cooption, regression, and growth in tumors mediated by angiopoietins and VEGF. Science 284:1994–1998
44. Oliner J, Min H, Leal J, Yu D, Rao S, You E et al (2004) Suppression of angiogenesis and tumor growth by selective inhibition of angiopoietin-2. Cancer Cell 6:507–516
45. Dumont DJ, Gradwohl G, Fong GH, Puri MC, Gertsenstein M, Auerbach A et al (1994) Dominant-negative and targeted null mutations in the endothelial receptor tyrosine kinase, tek, reveal a critical role in vasculogenesis of the embryo. Genes Dev 8(16):1897–1909
46. Sato TN, Tozawa Y, Deutsch U, Wolburg-Buchholz K, Fujiwara Y, Gendron-Maguire M et al (1995) Distinct roles of the receptor tyrosine kinases Tie-1 and Tie-2 in blood vessel formation. Nature 376(6535):70–74
47. Takakura N, Huang XL, Naruse T, Hamaguchi I, Dumont DJ, Yancopoulos GD et al (1998) Critical role of the TIE2 endothelial cell receptor in the development of definitive hematopoiesis. Immunity 9:677–686
48. Puri MC, Rossant J, Alitalo K, Bernstein A, Partanen J (1995) The receptor tyrosine kinase TIE is required for integrity and survival of vascular endothelial cells. EMBO J 14(23):5884–5891
49. Patan S (1998) TIE1 and TIE2 receptor tyrosine kinases inversely regulate embryonic angiogenesis by the mechanism of intussusceptive microvascular growth. Microvasc Res 56(1):1–21
50. Partanen J, Puri MC, Schwartz L, Fischer KD, Bernstein A, Rossant J (1996) Cell autonomous functions of the receptor tyrosine kinase TIE in a late phase of angiogenic capillary growth and endothelial cell survival during murine development. Development 122(10):3013–3021
51. Rodewald HR, Sato TN (1996) Tie1, a receptor tyrosine kinase essential for vascular endothelial cell integrity, is not critical for the development of hematopoietic cells. Oncogene 12:397–404
52. Qu X, Tompkins K, Batts LE, Puri M, Baldwin S (2010) Abnormal embryonic lymphatic vessel development in Tie1 hypomorphic mice. Development 137(8):1285–1295
53. Ward NL, Van Slyke P, Sturk C, Cruz M, Dumont DJ (2004) Angiopoietin 1 expression levels in the myocardium direct coronary vessel development. Dev Dyn 229:500–509
54. Suri C, McClain J, Thurston G, McDonald DM, Zhou H, Oldmixon EH et al (1998) Increased vascularization in mice overexpressing angiopoietin-1. Science 282:468–471

55. Reiss Y, Droste J, Heil M, Tribulova S, Schmidt MH, Schaper W et al (2007) Angiopoietin-2 impairs revascularization after limb ischemia. Circ Res 101:88–96
56. Fiedler U, Reiss Y, Scharpfenecker M, Grunow V, Koidl S, Thurston G et al (2006) Angiopoietin-2 sensitizes endothelial cells to TNF-alpha and has a crucial role in the induction of inflammation. Nat Med 12:235–239
57. Pitera JE, Woolf AS, Gale NW, Yancopoulos GD, Yuan HT (2004) Dysmorphogenesis of kidney cortical peritubular capillaries in angiopoietin-2-deficient mice. Am J Pathol 165:1895–1906
58. Hirakawa S, Hong YK, Harvey N, Schacht V, Matsuda K, Libermann T et al (2003) Identification of vascular lineage-specific genes by transcriptional profiling of isolated blood vascular and lymphatic endothelial cells. Am J Pathol 162:575–586
59. Morisada T, Oike Y, Yamada Y, Urano T, Akao M, Kubota Y et al (2005) Angiopoietin-1 promotes LYVE-1-positive lymphatic vessel formation. Blood 105:4649–4656
60. Tammela T, Saaristo A, Lohela M, Morisada T, Tornberg J, Norrmén C et al (2005) Angiopoietin-1 promotes lymphatic sprouting and hyperplasia. Blood 105:4642–4648
61. Dellinger M, Hunter R, Bernas M, Gale N, Yancopoulos G, Erickson R et al (2008) Defective remodeling and maturation of the lymphatic vasculature in Angiopoietin-2 deficient mice. Dev Biol 319:309–320
62. D'Amico G, Korhonen EA, Waltari M, Saharinen P, Laakkonen P, Alitalo K (2010) Loss of endothelial Tie1 receptor impairs lymphatic vessel development-brief report. Arterioscler Thromb Vasc Biol 30(2):207–209
63. Yuan L, Moyon D, Pardanaud L, Breant C, Karkkainen MJ, Alitalo K et al (2002) Abnormal lymphatic vessel development in neuropilin 2 mutant mice. Development 129:4797–4806
64. Kontos CD, Cha EH, York JD, Peters KG (2002) The endothelial receptor tyrosine kinase Tie1 activates phosphatidylinositol 3-kinase and Akt to inhibit apoptosis. Mol Cell Biol 22(6):1704–1713
65. Song SH, Kim KL, Lee KA, Suh W (2012) Tie regulates the Tie2 agonistic role of angiopoietin-2 in human lymphatic endothelial cells. Biochem Biophys Res Commun 419:281–286
66. Singh H, Hansen TM, Patel N, Brindle NP (2012) The molecular balance between receptor tyrosine kinases Tie1 and Tie2 is dynamically controlled by VEGF and TNFα and regulates angiopoietin signalling. PLoS One 7(1):e29319
67. Baumer S, Keller L, Holtmann A, Funke R, August B, Gamp A, Wolburg H, Wolburg-Buchholz K, Deutsch U, Vestweber D (2006) Vascular endothelial cell-specific phosphotyrosine phosphatase (VE-PTP) activity is required for blood vessel development. Blood 107:4754–4762
68. Dominguez MG, Hughes VC, Pan L, Simmons M, Daly C, Anderson K et al (2007) Vascular endothelial tyrosine phosphatase (VE-PTP)-null mice undergo vasculogenesis but die embryonically because of defects in angiogenesis. Proc Natl Acad Sci 104:3243–3248
69. Li Z, Huang H, Boland P, Dominguez MG, Burfeind P, Lai KM et al (2009) Embryonic stem cell tumor model reveals role of vascular endothelial receptor tyrosine phosphatase in regulating Tie2 pathway in tumor angiogenesis. Proc Natl Acad Sci 106:22399–22404
70. Winderlich M, Keller L, Cagna G, Broermann A, Kamenyeva O, Kiefer F et al (2009) VE-PTP controls blood vessel development by balancing Tie-2 activity. J Cell Biol 185:657–671
71. Huang L, Turck CW, Rao P, Peters KG (1995) GRB2 and SH-PTP2: potentially important endothelial signaling molecules downstream of the TEK/TIE2 receptor tyrosine kinase. Oncogene 11:2097–2103
72. Jones N, Master Z, Jones J, Bouchard D, Gunji Y, Sasaki H et al (1999) Identification of Tek/Tie2 binding partners. Binding to a multifunctional docking site mediates cell survival and migration. J Biol Chem 274:30896–30905
73. Audero E, Cascone I, Maniero F, Napione L, Arese M, Lanfrancone L et al (2004) Adaptor ShcA protein binds tyrosine kinase Tie2 receptor and regulates migration and sprouting but not survival of endothelial cells. J Biol Chem 279:13224–13233

74. Jones N, Dumont DJ (1998) The Tek/Tie2 receptor signals through a novel Dok-related docking protein, Dok-R. Oncogene 17:1097–1108
75. Jones N, Chen SH, Sturk C, Master Z, Tran J, Kerbel RS et al (2003) A unique autophosphorylation site on Tie2/Tek mediates Dok-R phosphotyrosine binding domain binding and function. Mol Cell Biol 23:2658–2668
76. Kim I, Ryu YS, Kwak HJ, Ahn SY, Oh JL, Yancopoulos GD et al (2002) EphB ligand, ephrinB2, suppresses the VEGF- and angiopoietin 1- induced Ras/mitogen-activated protein kinase pathway in venous endothelial cells. FASEB J 16:1126–1128
77. Harfouche R, Gratton JP, Yancopoulos GD, Noseda M, Karsan A, Hussain SN (2003) Angiopoietin-1 activates both anti- and proapoptotic mitogen-activated protein kinases. FASEB J 17:1523–1525
78. Saharinen P, Kerkelä K, Ekman N, Marron M, Brindle N, Lee GM et al (2005) Multiple angiopoietin recombinant proteins activate the Tie1 receptor tyrosine kinase and promote its interaction with Tie2. J Cell Biol 169:239–243
79. Yuan HT, Venkatesha S, Chan B, Deutsch U, Mammoto T, Sukhatme VP et al (2007) Activation of the orphan endothelial receptor Tie1 modifies Tie2-mediated intracellular signaling and cell survival. FASEB J 21(12):3171–3183
80. Eklund L, Olsen BR (2006) Tie receptors and their angiopoietin ligands are context-dependent regulators of vascular remodeling. Exp Cell Res 312:630–641

Chapter 9
VEGF Receptor Signaling in the Cardiac Lymphatics

Alexey Dashkevich, Karl B. Lemström, and Antti I. Nykänen

Abstract Since the discovery of angiogenic vascular endothelial growth factor (VEGF)-A in 1983 and lymphangiogenic VEGF-C in 1997, an increasing amount of knowledge has accumulated on the essential roles of VEGF ligands and receptors in physiological and pathological angiogenesis and lymphangiogenesis. We will review the properties of VEGF ligands and receptors concentrating on their lymphatic vessel effects first in noncardiac tissues and then in normal myocardium and cardiac disease. Tissue adaptation to several stimuli such as hypoxia, pathogen invasion, and inflammation often involves coordinated changes in both blood vessels and lymphatic vessels. As lymphatic vessels are involved in the initiation and resolution of inflammation and regulation of tissue edema, VEGF family members may have important roles in myocardial lymphatics and cardiac disease.

Keywords VEGF • Angiogenesis • Lymphangiogenesis • Cardiac lymphatics • Inflammation

---

A. Dashkevich (✉)
Transplantation Laboratory, University of Helsinki and Helsinki University Central Hospital, Helsinki, Finland
e-mail: alexey.dashkevich@helsinki.fi

K.B. Lemström • A.I. Nykänen
Transplantation Laboratory and Department of Cardiothoracic Surgery, University of Helsinki and Helsinki University General Hospital, Helsinki, Finland

G. Karunamuni (ed.), *The Cardiac Lymphatic System: An Overview*,
DOI 10.1007/978-1-4614-6774-8_9,

General Biology of the VEGF Family

VEGF-A (or VEGF) was first discovered as a tumor-secreted factor that induced vascular permeability and was therefore initially named as vascular permeability factor [1]. Subsequently, the VEGF family has expanded to a total of five mammalian ligands: placental growth factor (PlGF), VEGF-A, VEGF-B, VEGF-C, and VEGF-D [2–5].

The receptors for the VEGF family ligands include VEGFR-1, VEGFR-2, VEGFR-3, and neuropilins (NRP-1 and NRP-2). Upon binding of VEGF ligand dimers to their receptors, the corresponding receptors dimerize, phosphorylate tyrosine residues in the cytosol, and activate intracellular signaling pathways. The binding properties of VEGF ligands and the principal effects of VEGF receptors are summarized in Fig. 9.1. Heterodimerization of VEGF ligands and receptors, and splice variants and heparin-binding differences of ligands result in additional complexity and flexibility in VEGF ligand–receptor interactions [5]. Generally, the preferential expression of VEGFR-1 and VEGFR-2 in blood vascular endothelial cells (EC) and VEGFR-3 in lymphatic EC, and the binding properties of VEGF ligands to their corresponding receptors, underlies their angiogenic or lymphangiogenic predilection (Fig. 9.1). We first overview properties of the angiogenic VEGF ligands and receptors, and we then concentrate on the lymphangiogenic VEGF family members.

Angiogenic VEGF Members (VEGF-A, VEGF-B, PlGF, VEGFR-1, VEGFR-2, NRP)

VEGF-A binds to VEGFR-1, VEGFR-2, NRP-1, and NRP-2 and is the major regulator of the angiogenic switch. VEGF is produced by a variety of adult tissues, and vascular and inflammatory cells, and is induced by hypoxia, inflammation, and several growth factors [6–11]. VEGF-A elicits its effects mainly through VEGFR-2+ vascular endothelial cells (EC) and VEGFR-1+ monocytes and macrophages. VEGF induces EC migration, proliferation, and sprouting and results in angiogenesis. VEGF is essential for embryonic vascular development [12, 13]. The importance of VEGF in angiogenesis in adult is highlighted by the clinical success of anti-VEGF approaches in the treatment of cancer and eye disease [3]. However, the results of randomized clinical trials on VEGF-A-mediated therapeutic angiogenesis have been disappointing [14]. This may at least in part relate to the findings that the VEGF-A-induced new blood vessels are often leaky. VEGF-A is indeed intimately related to inflammation through increased vascular permeability and also through direct effects on VEGFR-1+ monocytes and macrophages [5].

PlGF was first detected in human placenta [15] and it binds to VEGFR-1 and NRP-1 [16]. PlGF is not essential for normal vascular development [3, 17], but it may enhance VEGF effects particularly in pathological conditions [18, 19], and it promotes cardiac hypertrophy [20]. PlGF has a clear inflammatory role as it activates both hematopoietic stem cells [1, 21] and monocytes and macrophages through VEGFR-1 [2, 22–24].

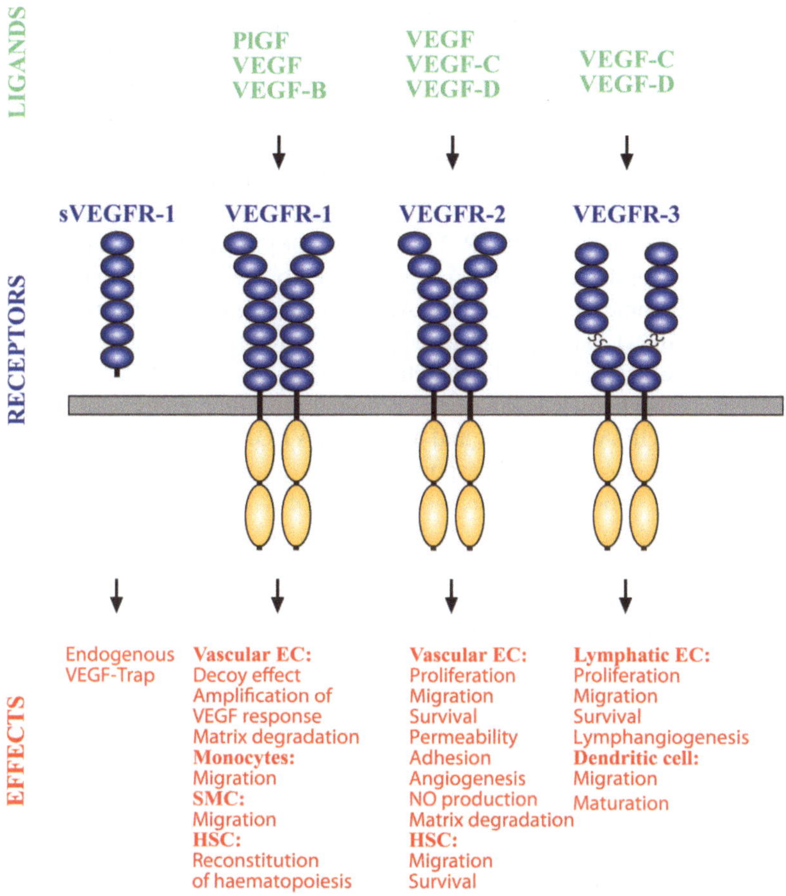

Fig. 9.1 VEGF ligands and receptors. VEGF ligands have different binding properties to VEGF receptors. VEGFR-1 modulates VEGF responses on endothelial cells and it is a chemotactic signal for monocytes and smooth muscle cells. VEGFR-1 also has a functional role in hematopoiesis, and the soluble sVEGFR-1 inhibits VEGF effects. VEGFR-2 elicits the main mitogenic and proinflammatory effects on vascular endothelial cells but has a functional role also in hematopoietic and vascular progenitor cells. VEGFR-3 signaling principally regulates the development and functionality of lymphatic endothelial cells and migration of antigen-presenting cells. *EC* endothelial cell; *HSC* hematopoietic stem cell; *NO* nitric oxide; *PlGF* placental growth factor; *SMC* smooth muscle cell; *VEGF* vascular endothelial growth factor; *VEGFR* VEGF receptor; *sVEGFR* soluble VEGFR (Modified from Nykänen A. Vascular growth factors and progenitor cells in cardiac allograft arteriosclerosis [dissertation]. Helsinki: University of Helsinki; 2007)

VEGF-B is expressed in skeletal muscle and heart, and also binds to VEGFR-1 and NRP-1 [3, 16, 25, 26]. Lack of VEGF-B does not impair vascular development, but it may yield to conducting defects and size reduction in the heart, and impaired recovery after myocardial ischemia [4, 27, 28]. VEGF-B is also involved in inflammatory angiogenesis [1, 29] and in cardiac arteriogenesis [5, 30] and hypertrophy [5, 31]. VEGF-B also has metabolic effects as it regulates endothelial lipid transfer and metabolism [6–11, 31, 32] as well as the development of diabetes [12, 13, 33].

VEGFR-1 (also known as fms-like tyrosine kinase, Flt-1) is a receptor for VEGF-A, VEGF-B, and PlGF [3, 16]. In addition to the membrane-anchored VEGFR-1, a soluble form of VEGFR-1 (sVEGFR-1) also exists [14, 34, 35]. VEGFR-1 is expressed in a variety of cells including EC, smooth muscle cells, monocytes and macrophages, and hematopoietic stem cells [3, 15–17, 36] and is upregulated by hypoxia-inducible factor-1 [18, 19, 37]. VEGFR-1 has high affinity for VEGF-A but low tyrosine kinase activity, and it has been viewed as a decoy receptor and a negative regulator of VEGFR-2 in ECs [20, 38]. Accordingly, VEGFR-1 deletion results in excessive endothelial progenitor cell proliferation, vascular disorganization, and embryonic lethality [39, 40]. Although the angiogenic effects of VEGFR-1 are subtle, VEGFR-1 may regulate arteriogenesis, pathological angiogenesis, myelomonocyte cell recruitment, and lipid metabolism [31–33, 41, 42].

VEGFR-2 (also known as kinase insert domain receptor, KDR/fetal liver kinase, Flk-1) binds VEGF-A, VEGF-C, and VEGF-D. VEGFR-2 is mainly expressed on vascular EC and is considered responsible for the majority of VEGF-A-induced angiogenic and permeability effects [3, 5]. VEGFR-2 expression is essential for embryonal hematopoiesis and vasculogenesis [43, 44]. In adults, VEGFR-2 expression is usually downregulated and presents only at sites of active angiogenesis such as wound healing, tumors, and after myocardial infarction [45, 46]. In myocardial infarction and sepsis, VEGFR-2 is a major regulator of vascular permeability and cardiac dysfunction [47, 48].

Neuropilin receptors NRP1 and NRP2 are involved in neuronal development and bind semaphorins [16]. Neuropilins also interact with VEGF signaling as NRP1 is a co-receptor for VEGFR-1 and VEGFR-2, and NRP2 for VEGFR-3 [16]. In the vascular system, NRP1 is expressed predominantly in arteries and potentiates the binding and activity of VEGF-A on VEGFR-2 [49]. Interaction of VEGFR-3 and its co-receptor NRP2 modulates the function of veins and lymphatic vessels [50, 51]. NRP1 also participates in lymphatic vessel and valve development through semaphorin interaction [52].

Although the VEGF ligands and receptors described above are considered mainly angiogenic, some overlap on lymphangiogenic effects have been described [53]. The reason for this overlap may be in part the fact that VEGF-A recruits macrophages that may in turn drive lymphangiogenesis [54].

Lymphangiogenic VEGF Members (VEGF-C and VEGF-D, VEGFR-3, NRP)

VEGF-C binds to VEGFR-2 and VEGFR-3 and is a major regulator in the development of lymphatic vasculature [2, 55–57]. VEGF-C is produced as a precursor protein that undergoes proteolytic modification [58]. It is produced in areas of active lymphangiogenesis in the embryo, and the expression is maintained in lung, heart, liver, and kidney in the adult [59]. VEGF-C is not activated by hypoxia, but is increased by serum and its components, platelet-derived growth factor, epidermal growth factor, and transforming growth factor-β [60]. VEGF-C mRNA is also upregulated by proinflammatory cytokines TNF-α, IL-1α, and IL-β [61]. Inflammatory cells such as macrophages, dendritic cells, and CD4+ T lymphocytes are a rich source of VEGF-C [54, 62–65].

Loss of one VEGF-C allele causes prominent lymphatic hypoplasia, whereas loss of both VEGF-C alleles results in embryonic lethality [66]. Similarly, inhibition of VEGF-C/D/R3 signaling in the adult regresses existing lymphatic vessels and results in lymphedema [67]. In contrast, VEGF-C overexpression results in lymphatic hyperplasia [2, 57] and in therapeutic lymphangiogenesis in lymphedema [68]. VEGF-C is upregulated in many cancers and participates in lymphatic metastasis of tumors [69, 70]. Interestingly, salt-induced hypertension is counteracted by macrophage VEGF-C upregulation and tissue lymphangiogenesis [71]. Another noteworthy finding suggests that, in addition to clear lymphatic effects, VEGF-C may also have angiogenic effects through VEGFR-2 binding, by regulating a subtype of vascular EC that express VEGFR-3 [72–74], and also by attracting VEGF-A-producing macrophages [75].

VEGF-C is intimately involved in many inflammatory conditions that involve lymphangiogenesis such as bacterial infection [62], rheumatoid arthritis [76], skin inflammation [77], and organ transplantation [64, 78, 79]. In addition to inducing lymphangiogenesis, VEGF-C modifies lymphatic vessel properties by, for example, upregulating CCL21—a chemokine that attracts CCR7+ tumor cells and dendritic cells [64, 80]. VEGF-C has also direct effects on VEGFR-3+ dendritic cells and induces their migration [81] and maturation [78]. VEGF-C thus modifies immune reactions through direct effects on lymphatic EC and also on macrophages and dendritic cells [64, 78, 81, 82]. In addition to participating in antigen-presenting cell traffic and the initiation of immune responses, VEGF-C-induced lymphangiogenesis may also balance tissue inflammation by promoting lymphatic drainage and the resolution of tissue inflammation [83–85].

Human VEGF-D binds to VEGFR-2 and VEGFR-3, whereas mouse VEGF-D binds only to VEGFR-3 [86]. Like VEGF-C, VEGF-D also undergoes proteolytical modification, is involved in lymphangiogenesis, and has angiogenic properties. In contrast to VEGF-C, VEGF-D is not essential for the development of lymphatic vessels [87]. Adenovirally delivered VEGF-D induces a potent angiogenic and lymphangiogenic response in skeletal muscle and is associated with elevated vascular permeability [88], but the relative effects on angiogenesis and lymphangiogenesis may depend on the tissue used [89]. VEGF-D is involved in the metastatic spread of cancer [90].

VEGFR-3 binds VEGF-C and VEGF-D, and it is a key regulator for lymphatic growth [2, 66]. VEGFR-3 signaling regulates cardiovascular development in embryos [91], but later in development and in adulthood, it more selectively regulates the growth and maintenance of lymphatic vessels [67]. VEGFR-3 may also have angiogenic effects in adults, since VEGFR-3 is expressed in stalk cells [72, 74], and VEGFR-3+ macrophages produce VEGF-A as well [75]. VEGFR-3 is defective in primary lymphedema [92, 93] and is induced in vascular malformations [94]. VEGFR-3 is also upregulated during lymphangiogenesis in cancer, wound healing, and inflammation [95]. Inflammation upregulates lymphatic EC VEGFR-3 expression trough the activation of inflammatory transcription factor NF-κB. This renders the VEGFR-3+ lymphatic ECs responsive to VEGF-C and VEGF-D [96]. In addition to lymphatic EC, VEGFR-3 is also expressed in macrophages and dendritic cells and regulates their migration [63, 81, 97].

Neuropilins are co-receptors for VEGFR-2 and VEGFR-3. NRP2 binds VEGF-C and VEGF-D, and is a co-receptor for VEGFR-3. In addition to its effects on veins, NRP2 also regulates lymphatic vessel sprouting [50, 51] and is upregulated in vascular malformations [94]. The other neuropilin NRP1 is also involved in lymphatic vessel maturation and valve formation, but this involves semaphorin and not VEGF-A [52].

Lymphatic-Specific VEGF Expression and Signaling in the Heart

VEGF Expression and Signaling in Cardiac Lymphatics During Embryogenesis and in Healthy Adult

The developing vasculature of the heart requires a variety of signals, including endothelial growth factors. The details of lymphatic development are very well described, mainly in noncardiac tissues of mouse, but the general features seem to be universal.

The first LECs differentiate from venous endothelial cells at midgestation, induced by VEGF-C [66, 98]. The LECs are distinguished by expression of specific molecules such as prospero-related homeodomain transcription factor (Prox1), vascular endothelial growth factor receptor-3 (VEGFR-3), the membrane glycoprotein podoplanin, and lymphatic vessel hyaluronan receptor-1 [66, 98, 99]. In mouse, starting from embryonic day (E) 9.5, a complex sequential activation of the transcription factor SOX18, Prox1, and the venous nuclear receptor COUP-TFII initiates the LEC differentiation program in the anterior cardinal vein [66, 98].

The paracrine secretion of VEGF-C is crucial for the further dorsolateral sprouting, migration, and survival of the first LECs and the formation of lymph sacs [66]. The VEGF-C co-receptor neuropilin-2 (NRP-2) and the Eph tyrosine kinase ligand ephrin B2 are required for efficient sprouting of lymphatic capillaries [51, 100]. The Notch1-Dll4 signaling pathway is essential for postnatal lymphatic development [101]. Interestingly, in adult tissues, lymphatic sprouting induced by VEGF-C is not restricted by Notch, whereas VEGF does not promote efficient lymphatic sprouting unless Notch signaling is inhibited [102].

According to the study of endothelial growth factor distribution in the human fetal heart [103], their localizations at different gestational ages are similar. Lymphatic vessels are only detected in the epicardial layer, and they are negative for VEGFR-1 but strongly positive for both VEGFR-2 and VEGFR-3. Very weak VEGFR-3 signals are also observed in some myocardial capillaries but not in the endocardium in 13- to 30-week fetuses. However, the VEGFR-3 expression seems to be downregulated in the blood vessels during the first trimester. Thus, although VEGFR-3 is needed for early cardiovascular development [104], it later serves a more specialized biological function mainly in lymphatic endothelia [103].

Only a few studies have described the pattern of VEGF receptor expression in the lymphatics of healthy adult heart [64, 105, 106]. In a study by Geissler et al. [105], human myocardial biopsies have been used for immunohistochemical stainings of VEGFR-3-positive lymphatics: the density of VEGFR-3-positive vessels was calculated to be around 50 per mm^2, and their average diameter was about 3 microns. As described for the healthy adult rat heart, the density of VEGFR-3-positive vessels is generally lower in the myocardium than in the epicardial area. In addition, vessel VEGFR-3 immunoreactivity colocalizes with LYVE-1 expression, although not all LYVE-1-positive vessels express VEGFR-3 [64].

In Heart Transplantation (Allorecognition and Rejection)

In transplantation, the transfer of antigen-presenting cells from vascularized allografts to secondary lymphoid organs—both spleen and lymph nodes—is critical for the priming of alloreactive T cells and the development of alloimmune responses [107, 108]. Lymphatic vessels provide easy access for inflammatory cells and their unilateral movement from peripheral tissues to secondary lymphoid organs. Thus, the lymphatic network forms a link between innate and adaptive immunity, thereby having extraordinary importance in a setting of transplantation.

A descriptive study of human patients undergoing heart transplantation shows that lymphatic endothelial markers undergo significant alterations after the transplantation, suggesting a significant change in lymphatic endothelial phenotype. Furthermore, episodes of acute allograft rejection seem to be associated with a significantly lower density of VEGFR-3-positive lymphatics after heart transplantation [106]. A recent experimental study [64] provides detailed information on lymphatic behavior in rejecting hearts: acute rejection decreases the epicardial lymphatic vessel density and chronic rejection doubles the myocardial lymphatic vessel density. Importantly, lymphangiogenesis in transplanted organs may not be only a secondary effect of chronic inflammation. Instead, lymphatic vessels also appear to have a regulatory role in the initiation of alloimmune reactions. VEGFR-3 inhibition decreases dendritic cell recruitment to the spleen and the development of the subsequent alloimmune response, thus improving cardiac allograft survival. In addition, treatment with neutralizing monoclonal VEGFR-3 antibodies decreases allograft inflammation and the development of arteriosclerosis in a chronic rejection model. According to the study, it appears possible that VEGFR-3 inhibition has direct effects on dendritic cell migration. VEGFR-3 also seems to regulate leukocyte traffic and alloimmunity through direct effects on allograft VEGFR-3-positive lymphatic vessels by upregulating allograft CCL21 production [64]. The range of currently known directions for VEGF-C signaling in a setting of transplantation is represented in Fig. 9.2.

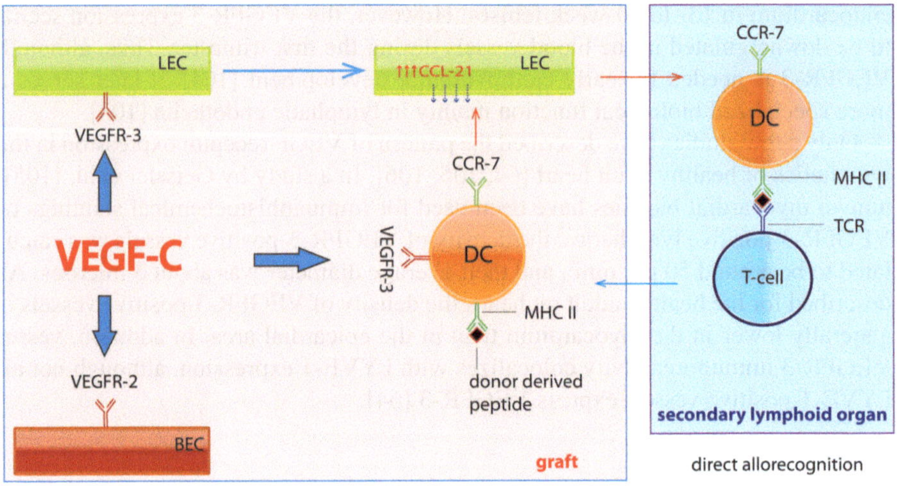

Fig. 9.2 VEGF-C–VEGFR-3 signaling in a setting of transplantation. VEGF-C modifies lymphatic vessel properties by upregulating CCL21—a chemokine that attracts CCR7+ dendritic cells. VEGF-C also has direct effects on VEGFR-3+ dendritic cells and induces their maturation and unilateral migration through the lymphatic network to secondary lymphoid organs. Thus, VEGF-C plays an important role in the initiation of direct alloimmune recognition through direct effects on lymphatic endothelial cells and antigen-presenting cells. VEGF-C may also have angiogenic effects through VEGFR-2 binding; its role in transplantation remains unclear. *BEC* blood endothelial cell; *CCL-21* chemokine ligand 21; *LEC* lymphatic endothelial cell; *CCR-7* C-C chemokine receptor type 7; *DC* dendritic cell; *MHC II* major histocompatibility complex class II; *VEGFR-2* VEGF receptor 2; *VEGFR-3* VEGF receptor 3; *TCR* T cell receptor

Thus, VEGF-C/VEGFR-3 signaling seems to have important effects on proximal events in cardiac allograft alloimmunity and arteriosclerosis. Therefore, VEGFR-3 inhibition could be used as a novel lymphatic vessel-targeted immunomodulatory therapy to regulate alloimmune activation after solid organ transplantation. Further studies that particularly describe the mechanistic role of lymphatic vessel activation in the ischemia-reperfusion injury and allograft rejection are needed.

In Myocardial Infarction

In the human heart, myocardial remodeling after myocardial infarction (MI) results in scar formation through several sequential stages of myocardial necrosis, granulation, and fibrosis [109]. The viable cardiomyocytes around the lesion express cytoprotective proteins and cytokines which facilitate the healing of the affected lesion [110, 111]. VEGF is critical for angiogenesis in the healing area [112]: it is promptly expressed in the surviving cardiomyocytes around the infracted lesion after the onset of MI, and angiogenesis in the lesion begins at 4–5 h and continues up to day 90 [110].

A study by Ishikawa et al. [113] demonstrates that lymphatic vessels are not detected in stages with coagulation necrosis, but a few lymphatics first appear in the peripheral region adjacent to viable myocytes in the early granulation period. Lymphatic density subsequently increases in the mature granulation period and is thereafter maintained during scar formation. After the onset of myocardial infarction, lymphangiogenesis lags behind blood vessel angiogenesis, whereas VEGF-C is expressed in the cardiomyocytes around the lesion at all stages of myocardial remodeling. The results of this study suggest that during the entire process of myocardial infarction healing, blood vessels supply the blood and nutrients mainly during the granulation period, but lymphatics participate mainly in fibrosis maturation and scar formation through the drainage of excessive proteins and fluid, probably mediated by VEGF-C.

In Atherosclerosis and Degenerative Valve Disease

Recently there has been an emerging body of evidence linking lymphangiogenesis to atherosclerosis. Lymphatic vessels are found at sites of atherosclerosis, which is associated with inflammation and lipid accumulation in arterial walls [114, 115]. Arterial smooth muscle cells constitutively produce lymphangiogenic factors, and lymphatic vessels are present in the adventitia of arteries adjacent to small blood vessels, called the vasa vasorum, which are expanded in atherosclerotic plaques [116, 117]. However, according to the study of Nakano et al. [117], impaired lymphangiogenesis may contribute to plaque progression. Here, VEGF-A and VEGF-C might synergistically contribute to angiogenesis in coronary atherosclerotic plaques. So, in atherosclerotic lesions, the imbalance of angiogenesis and lymphangiogenesis in favor of angiogenesis seems to contribute to sustained inflammatory reactions during human coronary atherogenesis.

In atherosclerosis, lymphatic vessels might have an important role in the efflux of interstitial fluids, fats, and inflammatory cells from the wall of the coronary artery, slowing down the development of atherosclerotic lesions. It remains to be investigated whether therapeutic targeting of lymphangiogenesis might reveal an antiatherosclerotic tool.

Aortic valve stenosis (AS) is another degenerative disease of the heart, where the pathogenesis is linked to lymphangiogenic factors. Pathological features of AS are calcification [118], extracellular matrix remodeling, and valvular accumulation of lipids and inflammatory cells [119]. In contrast to normal avascular aortic valves, stenotic aortic valves are vascularized [120]. Importantly, neovessels may contribute to the progression of AS by facilitating the entry of inflammatory cells and lipids into the leaflets [121]. The study of Syväranta et al. [122] demonstrates that lymphangiogenesis is a part of the pathogenesis of AS and shows that myofibroblasts and endothelial cells are responsible for the valvular production of lymphangiogenic growth factors VEGF-C and VEGF-D.

Furthermore, mast-cell-derived compounds degrade VEGF-C, suggesting their anti-lymphangiogenic potential. Similar to atherosclerotic lesions, in AS, the balance between angiogenesis and lymphangiogenesis may disadvantageously favor the accumulation of inflammatory cells and lipids into the lesions, thus possibly leading to disease progression.

In Heart Failure

The two most frequent causes of terminal heart failure are dilated (DCM) and ischemic (ICM) cardiomyopathy. The investigation of microvascular structures in cardiac remodeling has mostly been limited to the sequela of myocardial ischemia and infarction rather than in terminal heart failure. Nevertheless, the role of lymphatic system in the failing myocardium might be of increased importance, since the hemodynamics, fluid, and pressure balance are severely damaged. However, there are barely any publications available about lymphatics in the failing heart, except for several descriptional reports.

The study of Aharinejad et al. [123] provides evidence that VEGF-C mRNA levels are upregulated in both forms of cardiomyopathy and that after cardiac transplantation, these mRNA levels returned to the baseline level of nonfailing cardiac tissues in DCM or decreased even below the baseline level in ICM. A further study [124] describes the distribution of several lymphatic and blood markers, including VEGFRs, in DCM and ICM, comparing them to nonfailing hearts.

Whether any therapeutic manipulation of lymphatic function could improve impaired myocardial function by draining the failing myocardium remains so far an exciting speculation.

In Inflammation

In adulthood, lymphangiogenesis and elevated VEGFR-3 expression coincide with various inflammatory conditions including cancer [125], wound healing [126], and chronic inflammatory diseases. Increased lymphatic vessel density has been documented in chronic airway infection [62], psoriasis [127], and arthritis [76]. VEGF-C and VEGF-D are elevated during inflammation, being produced by a variety of cells residing at inflamed sites, including macrophages [62, 128, 129], dendritic cells and neutrophils [62], mast cells and fibroblasts [130], and tumor cells [129].

Generally, the role of lymphatic activation and lymphangiogenesis in inflammatory settings is considered to be positive. It facilitates the resolution of tissue edema and enhances immune responses by promoting macrophage and dendritic cell mobilization [56, 62]. The lymphatic vascular system and the molecular

Table 9.1 Overview of lymphatic-specific VEGF expression and signaling in the heart during embryogenesis, in healthy adult, and in various disease states of the heart

	VEGF member	Expression and signaling pattern
Embryogenesis	VEGF-C	• Induces the differentiation of first LECs from venous endothelial cells [66, 98] • Crucial for further dorsolateral sprouting, migration and survival of the first LECs, and the formation of lymph sacs [66]
	VEGFR-2	Expressed already on the first LECs [103]
	VEGFR-3	Expressed already on the first LECs, as well as on cardiac blood vessels during the first trimester [103]
Healthy adult	VEGFR-2	Expressed in a small number of lymphatic vessels (no precise data available)
	VEGFR-3	Expressed in a considerable number of lymphatic vessels (no precise data available) [64, 105, 106]
Heart transplantation	VEGF-C	• Modifies lymphatic vessel properties by upregulating CCL21 [64] • Induces maturation and migration of dendritic cells [78]
	VEGFR-3	• Important downstream target for VEGF-C effects [64] • Changes in densities of VEGFR-3+ vessels accompany episodes of allograft rejection [106]
Myocardial infarction	VEGF-C	Mediates lymphatic participation in fibrosis maturation and scar formation through the drainage of excessive proteins and fluids [113]
Atherosclerosis	VEGF-C	Contributes with VEGF-A to angiogenesis in coronary plaque, leading to imbalance of angio- and lymphangiogenesis, thus sustaining inflammatory reaction during atherogenesis [117]
Aortic stenosis	VEGF-C	Is degraded in diseased valve leaflets by mast-cell compounds, which leads to imbalance of angio- and lymphangiogenesis and favors the accumulation of inflammatory cells and lipids (disease progression) [122]
Inflammation		Although lymphatic growth accompanies infective heart diseases, chronic inflammation, infarction, etc., the role of lymphatic-specific VEGF signaling in these processes has not been studied
Heart failure		Only controversial purely descriptive data is available [123, 124]

pathways regulating inflammatory responses are intimately associated. Lymphatic vessels react to tissue inflammation with morphological changes in lymphatic endothelial cell phenotype (such as overexpression of adhesion molecules [131] or VEGFR-3 [96]), induction of proinflammatory cytokines production, as well as chronologically delayed increase of lymphatic vessel density (lymphangiogenesis) [96].

LECs at least in some tissues constitutively express NF-κB [132]. Activation of the NF-κB pathway in LECs upregulates Prox1 and VEGFR-3, which renders the lymphatic vessels more sensitive to VEGF-C and VEGF-D produced by leukocytes [96]. VEGFR-3 signaling is activated upon binding of vascular endothelial growth factor-C (VEGF-C) or the related factor, VEGF-D [133]. NF-κB is activated, for example, downstream of Toll-like receptor 4 binding to lipopolysaccharide in the LECs, thus inducing the activation and production of leukocyte chemoattractants such as CCL2, CCL5, and CX3CL1, which in turn promotes leukocyte homing to the lymphatic vessels and eventually to the draining lymph node [134].

Inflamed lymphatic endothelium promotes the exit of leukocytes from tissue to afferent lymph through newly induced expression of the adhesion molecules ICAM-1 and VCAM-1, which were previously thought to be specific for blood vessel transmigration [131].

Furthermore, the studies of anatomical distribution for the cardiac lymphatics in diseased heart demonstrate increased lymphatic densities in infective endocarditis, myocarditis, and progressive atherosclerosis. Thus, lymphatic growth accompanies chronic inflammation, tissue degeneration, infarction, calcification, and formation of connective tissue [115]. Although not investigated yet for these disease states, the role of lymphatic-specific VEGF signaling seems to be crucial and warrants further analysis and a search for potential therapeutic targeting. The properties of lymphatic-specific VEGF expression and signaling in the heart are briefly summarized in Table 9.1.

Summary

While the critical role of angiogenic VEGF family members in cardiovascular development and disease is already appreciated, the involvement of lymphangiogenic VEGF ligands and receptors, and cardiac lymphatics, in cardiac physiology and pathology is only starting to unfold. With the rapid development of lymphatic vessel markers, better understanding of basic lymphatic vessel biology, and the use of novel genetic and pharmacological tools to activate or inhibit lymphangiogenic VEGF-C/D/R3 signaling, future studies may reveal novel lymphatic-targeted therapeutic strategies in ischemic, degenerative, and inflammatory heart disease.

References

1. Senger DR, Galli SJ, Dvorak AM, Perruzzi CA, Harvey VS, Dvorak HF (1983) Tumor cells secrete a vascular permeability factor that promotes accumulation of ascites fluid. Science 219:983–985
2. Jeltsch M, Kaipainen A, Joukov V, Meng X, Lakso M, Rauvala H, Swartz M, Fukumura D, Jain RK, Alitalo K (1997) Hyperplasia of lymphatic vessels in VEGF-C transgenic mice. Science 276:1423–1425
3. Carmeliet P, Jain RK (2011) Molecular mechanisms and clinical applications of angiogenesis. Nature 473:298–307

4. Alitalo A, Detmar M (2012) Interaction of tumor cells and lymphatic vessels in cancer progression. Oncogene 31:4499–4508
5. Koch S, Claesson-Welsh L (2012) Signal transduction by vascular endothelial growth factor receptors. Cold Spring Harb Perspect Med 2:a006502
6. Berse B, Brown LF, Van de Water L, Dvorak HF, Senger DR (1992) Vascular permeability factor (vascular endothelial growth factor) gene is expressed differentially in normal tissues, macrophages, and tumors. Mol Biol Cell 3:211–220
7. Ferrara N, Winer J, Burton T (1990) Aortic smooth muscle cells express and secrete vascular endothelial growth factor. Growth Factors 5:141–148
8. Freeman MR, Schneck FX, Gagnon ML, Corless C, Soker S, Niknejad K et al (1995) Peripheral blood T lymphocytes and lymphocytes infiltrating human cancers express vascular endothelial growth factor: a potential role for T cells in angiogenesis. Cancer Res 55:4140–4145
9. McCourt M, Wang JH, Sookhai S, Redmond HP (1999) Proinflammatory mediators stimulate neutrophil-directed angiogenesis. Arch Surg 134:1325–1331
10. Melter M, Reinders ME, Sho M, Pal S, Geehan C, Denton MD et al (2000) Ligation of CD40 induces the expression of vascular endothelial growth factor by endothelial cells and monocytes and promotes angiogenesis in vivo. Blood 96:3801–3808
11. Tomizawa S, Maruyama K, Nagasawa N, Suzuki S, Kuroume T (1984) Studies of vascular permeability factor derived from T lymphocytes and inhibitory effect of plasma on its production in minimal change nephrotic syndrome. Nephron 41:157–160
12. Carmeliet P, Ferreira V, Breier G, Pollefeyt S, Kieckens L, Gertsenstein M et al (1996) Abnormal blood vessel development and lethality in embryos lacking a single VEGF allele. Nature 380:435–439
13. Ferrara N, Carver-Moore K, Chen H, Dowd M, Lu L, O'Shea KS et al (1996) Heterozygous embryonic lethality induced by targeted inactivation of the VEGF gene. Nature 380:439–442
14. Giacca M, Zacchigna S (2012) VEGF gene therapy: therapeutic angiogenesis in the clinic and beyond. Gene Ther 19:622–629
15. Maglione D, Guerriero V, Viglietto GG, Delli-Bovi PP, Persico MGM (1991) Isolation of a human placenta cDNA coding for a protein related to the vascular permeability factor. Proc Natl Acad Sci USA 88:9267–9271
16. Koch S (2012) Neuropilin signalling in angiogenesis. Biochem Soc Trans 40:20–25
17. Gigante B, Tarsitano M, Cimini V, De Falco S, Persico MG (2003) Placenta growth factor is not required for exercise-induced angiogenesis. Angiogenesis 7:277–284
18. Carmeliet P, Moons L, Luttun A, Vincenti V, Compernolle V, De Mol M et al (2001) Synergism between vascular endothelial growth factor and placental growth factor contributes to angiogenesis and plasma extravasation in pathological conditions. Nat Med 7:575–583
19. Park JE, Chen HH, Winer J, Houck KA, Ferrara N (1994) Placenta growth factor. Potentiation of vascular endothelial growth factor bioactivity, in vitro and in vivo, and high affinity binding to Flt-1 but not to Flk-1/KDR. J Biol Chem 269:25646–25654
20. Accornero F, van Berlo JH, Benard MJ, Lorenz JN, Carmeliet P, Molkentin JD (2011) Placental growth factor regulates cardiac adaptation and hypertrophy through a paracrine mechanism. Circ Res 109:272–280
21. Rafii S, Avecilla S, Shmelkov S, Shido K, Tejada R, Moore MAS et al (2003) Angiogenic factors reconstitute hematopoiesis by recruiting stem cells from bone marrow microenvironment. Ann N Y Academy Sci 996:49–60
22. Adini A, Kornaga T, Firoozbakht F, Benjamin LE (2002) Placental growth factor is a survival factor for tumor endothelial cells and macrophages. Cancer Res 62:2749–2752
23. Barleon B, Sozzani S, Zhou D, Weich HA, Mantovani A, Marmé D (1996) Migration of human monocytes in response to vascular endothelial growth factor (VEGF) is mediated via the VEGF receptor flt-1. Blood 87:3336–3343

24. Clauss M, Weich H, Breier G, Knies U, Röckl W, Waltenberger J et al (1996) The vascular endothelial growth factor receptor Flt-1 mediates biological activities: implications for a functional role of placenta growth factor in monocyte activation and chemotaxis. J Biol Chem 271:17629–17634
25. Olofsson B, Pajusola K, von Euler G, Chilov D, Alitalo K, Eriksson U (1996) Genomic organization of the mouse and human genes for vascular endothelial growth factor B (VEGF-B) and characterization of a second splice isoform. J Biol Chem 271:19310–19317
26. Olofsson B, Pajusola K, Kaipainen A, von Euler G, Joukov V, Saksela O et al (1996) Vascular endothelial growth factor B, a novel growth factor for endothelial cells. Proc Natl Acad Sci USA 93:2576–2581
27. Aase K, von Euler G, Li X, Pontén A, Thorén P, Cao R et al (2001) Vascular endothelial growth factor-B-deficient mice display an atrial conduction defect. Circulation 104:358–364
28. Bellomo D, Headrick JP, Silins GU, Paterson CA, Thomas PS, Gartside M et al (2000) Mice lacking the vascular endothelial growth factor-B gene (Vegfb) have smaller hearts, dysfunctional coronary vasculature, and impaired recovery from cardiac ischemia. Circ Res 86:29–35
29. Mould AW, Tonks ID, Cahill MM, Pettit AR, Thomas R, Hayward NK et al (2003) Vegfb gene knockout mice display reduced pathology and synovial angiogenesis in both antigen-induced and collagen-induced models of arthritis. Arthritis Rheum 48:2660–2669
30. Bry M, Kivelä R, Holopainen T, Anisimov A, Tammela T, Soronen J et al (2010) Vascular endothelial growth factor-B acts as a coronary growth factor in transgenic rats without inducing angiogenesis, vascular leak, or inflammation. Circulation 122:1725–1733
31. Kärpänen T, Bry M, Ollila HM, Seppänen-Laakso T, Liimatta E, Leskinen H et al (2008) Overexpression of vascular endothelial growth factor-B in mouse heart alters cardiac lipid metabolism and induces myocardial hypertrophy. Circ Res 103:1018–1026
32. Hagberg CE, Falkevall A, Wang X, Larsson E, Huusko J, Nilsson I et al (2010) Vascular endothelial growth factor B controls endothelial fatty acid uptake. Nature 464:917–921
33. Hagberg CE, Mehlem A, Falkevall A, Muhl L, Fam BC, Ortsäter H et al (2012) Targeting VEGF-B as a novel treatment for insulin resistance and type 2 diabetes. Nature 490:426–430
34. Ambati BK, Nozaki M, Singh N, Takeda A, Jani PD, Suthar T et al (2006) Corneal avascularity is due to soluble VEGF receptor-1. Nature 443:993–997
35. Kendall RL, Thomas KA (1993) Inhibition of vascular endothelial cell growth factor activity by an endogenously encoded soluble receptor. Proc Natl Acad Sci USA 90:10705–10709
36. Couper LL, Bryant SR, Eldrup-Jørgensen J, Bredenberg CE, Lindner V (1997) Vascular endothelial growth factor increases the mitogenic response to fibroblast growth factor-2 in vascular smooth muscle cells in vivo via expression of fms-like tyrosine kinase-1. Circ Res 81:932–939
37. Gerber HP, Condorelli F, Park J, Ferrara N (1997) Differential transcriptional regulation of the two vascular endothelial growth factor receptor genes. Flt-1, but not Flk-1/KDR, is up-regulated by hypoxia. J Biol Chem 272:23659–23667
38. Hiratsuka S, Minowa O, Kuno J, Noda T, Shibuya M (1998) Flt-1 lacking the tyrosine kinase domain is sufficient for normal development and angiogenesis in mice. Proc Natl Acad Sci USA 95:9349–9354
39. Fong GH, Rossant J, Gertsenstein M, Breitman ML (1995) Role of the Flt-1 receptor tyrosine kinase in regulating the assembly of vascular endothelium. Nature 376:66–70
40. Fong GH, Zhang L, Bryce DM, Peng J (1999) Increased hemangioblast commitment, not vascular disorganization, is the primary defect in flt-1 knock-out mice. Development 126:3015–3025
41. Luttun A, Tjwa M, Moons L, Wu Y, Angelillo-Scherrer A, Liao F et al (2002) Revascularization of ischemic tissues by PlGF treatment, and inhibition of tumor angiogenesis, arthritis and atherosclerosis by anti-Flt1. Nat Med 8:831–840
42. Pipp F, Heil M, Issbrücker K, Ziegelhoeffer T, Martin S, van den Heuvel J et al (2003) VEGFR-1-selective VEGF homologue PlGF is arteriogenic: evidence for a monocyte-mediated mechanism. Circ Res 92:378–385

43. Kaipainen A, Korhonen J, Pajusola K, Aprelikova O, Persico MG, Terman BI et al (1993) The related FLT4, FLT1, and KDR receptor tyrosine kinases show distinct expression patterns in human fetal endothelial cells. J Exp Med 178:2077–2088
44. Shalaby F, Rossant J, Yamaguchi TP, Gertsenstein M, Wu XF, Breitman ML et al (1995) Failure of blood-island formation and vasculogenesis in Flk-1-deficient mice. Nature 376:62–66
45. Li J, Brown LF, Hibberd MG, Grossman JD, Morgan JP, Simons M (1996) VEGF, flk-1, and flt-1 expression in a rat myocardial infarction model of angiogenesis. Am J Physiol 270:H1803–H1811
46. Shibuya M (1994) Role of VEGF-flt receptor system in normal and tumor angiogenesis. Adv Cancer Res 67:281–316
47. Weis S, Shintani S, Weber A, Kirchmair R, Wood M, Cravens A et al (2004) Src blockade stabilizes a Flk/cadherin complex, reducing edema and tissue injury following myocardial infarction. J Clin Invest 113:885–894
48. Yano K, Liaw PC, Mullington JM, Shih S-C, Okada H, Bodyak N et al (2006) Vascular endothelial growth factor is an important determinant of sepsis morbidity and mortality. J Exp Med 203:1447–1458
49. Soker S, Takashima S, Miao HQ, Neufeld G, Klagsbrun M (1998) Neuropilin-1 is expressed by endothelial and tumor cells as an isoform-specific receptor for vascular endothelial growth factor. Cell 92:735–745
50. Kärpänen T, Heckman CA, Keskitalo S, Jeltsch M, Ollila H, Neufeld G et al (2006) Functional interaction of VEGF-C and VEGF-D with neuropilin receptors. FASEB J 20:1462–1472
51. Xu Y, Yuan L, Mak J, Pardanaud L, Caunt M, Kasman I et al (2010) Neuropilin-2 mediates VEGF-C-induced lymphatic sprouting together with VEGFR3. J Cell Biol 188:115–130
52. Jurisic G, Hajjami HM, Karaman S, Ochsenbein AM, Alitalo A, Siddiqui SS et al (2012) An unexpected role of semaphorin3a-neuropilin-1 signaling in lymphatic vessel maturation and valve formation. Circ Res 111:426–436
53. Nagy JA, Vasile E, Feng D, Sundberg C, Brown LF, Detmar MJ et al (2002) Vascular permeability factor/vascular endothelial growth factor induces lymphangiogenesis as well as angiogenesis. J Exp Med 196:1497–1506
54. Cursiefen C, Chen L, Borges LP, Jackson D, Cao J, Radziejewski C et al (2004) VEGF-A stimulates lymphangiogenesis and hemangiogenesis in inflammatory neovascularization via macrophage recruitment. J Clin Invest 113:1040–1050
55. Alitalo K (2011) The lymphatic vasculature in disease. Nat Med 17:1371–1380
56. Kim H, Kataru RP, Koh GY (2012) Regulation and implications of inflammatory lymphangiogenesis. Trends Immunol 33:350–356
57. Oh SJ, Jeltsch MM, Birkenhäger R, McCarthy JE, Weich HA, Christ B et al (1997) VEGF and VEGF-C: specific induction of angiogenesis and lymphangiogenesis in the differentiated avian chorioallantoic membrane. Dev Biol 188:14–24
58. Joukov V, Pajusola K, Kaipainen A, Chilov D, Lahtinen I, Kukk E et al (1996) A novel vascular endothelial growth factor, VEGF-C, is a ligand for the Flt4 (VEGFR-3) and KDR (VEGFR-2) receptor tyrosine kinases. EMBO J 15:290–298
59. Kukk E, Lymboussaki A, Taira S, Kaipainen A, Jeltsch M, Joukov V et al (1996) VEGF-C receptor binding and pattern of expression with VEGFR-3 suggests a role in lymphatic vascular development. Development 122:3829–3837
60. Enholm B, Paavonen K, Ristimäki A, Kumar V, Gunji Y, Klefstrom J et al (1997) Comparison of VEGF, VEGF-B, VEGF-C and Ang-1 mRNA regulation by serum, growth factors, oncoproteins and hypoxia. Oncogene 14:2475–2483
61. Ristimäki A, Narko K, Enholm B, Joukov V, Alitalo K (1998) Proinflammatory cytokines regulate expression of the lymphatic endothelial mitogen vascular endothelial growth factor-C. J Biol Chem 273:8413–8418
62. Baluk P, Tammela T, Ator E, Lyubynska N, Achen MG, Hicklin DJ et al (2005) Pathogenesis of persistent lymphatic vessel hyperplasia in chronic airway inflammation. J Clin Invest 115:247–257

63. Hamrah P, Chen L, Zhang Q, Dana MR (2003) Novel expression of vascular endothelial growth Ffctor receptor (VEGFR)-3 and VEGF-C on corneal dendritic cells. Am J Pathol 163:12–22
64. Nykanen AI, Sandelin H, Krebs R, Keränen MAI, Tuuminen R, Kärpänen T et al (2010) Targeting lymphatic vessel activation and CCL21 production by vascular endothelial growth factor receptor-3 inhibition has novel immunomodulatory and antiarteriosclerotic effects in cardiac allografts. Circulation 121:1413–1422
65. Schoppmann SF, Birner P, Stöckl J, Kalt R, Ullrich R, Caucig C et al (2002) Tumor-associated macrophages express lymphatic endothelial growth factors and are related to peritumoral lymphangiogenesis. Am J Pathol 161(3):947–956
66. Karkkainen MJ, Haiko P, Sainio K, Partanen J, Taipale J, Petrova TV et al (2004) Vascular endothelial growth factor C is required for sprouting of the first lymphatic vessels from embryonic veins. Nat Immunol 5:74–80
67. Makinen T, Jussila L, Veikkola T, Karpanen T, Kettunen MI, Pulkkanen KJ et al (2001) Inhibition of lymphangiogenesis with resulting lymphedema in transgenic mice expressing soluble VEGF receptor-3. Nat Med 7:199–205
68. Szuba A, Skobe M, Karkkainen MJ, Shin WS, Beynet DP, Rockson NB et al (2002) Therapeutic lymphangiogenesis with human recombinant VEGF-C. FASEB J 16:1985–1987
69. Mandriota SJ, Jussila L, Jeltsch M, Compagni A, Baetens D, Prevo R et al (2001) Vascular endothelial growth factor-C-mediated lymphangiogenesis promotes tumour metastasis. EMBO J 20:672–682
70. Skobe M, Hawighorst T, Jackson DG, Prevo R, Janes L, Velasco P et al (2001) Induction of tumor lymphangiogenesis by VEGF-C promotes breast cancer metastasis. Nat Med 7:192–198
71. Machnik A, Neuhofer W, Jantsch J, Dahlmann A, Tammela T, Machura K et al (2009) Macrophages regulate salt-dependent volume and blood pressure by a vascular endothelial growth factor-C-dependent buffering mechanism. Nat Med 15:545–552
72. Benedito R, Rocha SF, Woeste M, Zamykal M, Radtke F, Casanovas O et al (2012) Notch-dependent VEGFR3 upregulation allows angiogenesis without VEGF-VEGFR2 signalling. Nature 484(7392):110–114
73. Saaristo A, Veikkola T, Enholm B, Hytönen M, Arola J, Pajusola K et al (2002) Adenoviral VEGF-C overexpression induces blood vessel enlargement, tortuosity, and leakiness but no sprouting angiogenesis in the skin or mucous membranes. FASEB J 16:1041–1049
74. Tammela T, Zarkada G, Nurmi H, Jakobsson L, Heinolainen K, Tvorogov D et al (2011) VEGFR-3 controls tip to stalk conversion at vessel fusion sites by reinforcing Notch signalling. Nat Cell Biol 13:1202–1213
75. Chung ES, Chauhan SK, Jin Y, Nakao S, Hafezi-Moghadam A, Van Rooijen N et al (2009) Contribution of macrophages to angiogenesis induced by vascular endothelial growth factor receptor-3-specific ligands. Am J Pathol 175:9–19
76. Zhang Q, Lu Y, Proulx ST, Guo R, Yao Z, Schwarz EM et al (2007) Increased lymphangiogenesis in joints of mice with inflammatory arthritis. Arthritis Res Ther 9:118
77. Shi VY, Bao L, Chan LS (2012) Inflammation-driven dermal lymphangiogenesis in atopic dermatitis is associated with CD11b+ macrophage recruitment and VEGF-C up-regulation in the IL-4-transgenic mouse model. Microcirculation 19:567–579
78. Hajrasouliha AR, Funaki T, Sadrai Z, Hattori T, Chauhan SK, Dana R (2012) Vascular endothelial growth factor-C promotes alloimmunity by amplifying antigen presenting cell maturation and lymphangiogenesis. Invest Ophthalmol Vis Sci 53(3):1244–1250
79. Krebs R, Tikkanen JM, Ropponen JO, Jeltsch M, Jokinen JJ, Herttuala SY et al (2012) Critical role of VEGF-C/VEGFR-3 signaling in innate and adaptive immune responses in experimental obliterative bronchiolitis. Am J Pathol 181:1607–1620
80. Issa A, Le TX, Shoushtari AN, Shields JD, Swartz MA (2008) Vascular endothelial growth factor-C and C-C chemokine receptor 7 in tumor cell-lymphatic cross-talk promote invasive phenotype. Cancer Res 69:349–357

81. Chen L, Hamrah P, Cursiefen C, Zhang Q, Pytowski B, Streilein JW et al (2004) Vascular endothelial growth factor receptor-3 mediates induction of corneal alloimmunity. Nat Med 10:813–815
82. Skobe M, Hamberg LM, Hawighorst T, Schirner M, Wolf GL, Alitalo K et al (2001) Concurrent induction of lymphangiogenesis, angiogenesis, and macrophage recruitment by vascular endothelial growth factor-C in melanoma. Am J Pathol 159:893–903
83. Huggenberger R, Siddiqui SS, Brander D, Ullmann S, Zimmermann K, Antsiferova M et al (2011) An important role of lymphatic vessel activation in limiting acute inflammation. Blood 117:4667–4678
84. Huggenberger R, Ullmann S, Proulx ST, Pytowski B, Alitalo K, Detmar M (2010) Stimulation of lymphangiogenesis via VEGFR-3 inhibits chronic skin inflammation. J Exp Med 207:2255–2269
85. Zhou Q, Guo R, Wood R, Boyce BF, Liang Q, Wang YJ et al (2011) Vascular endothelial growth factor C attenuates joint damage in chronic inflammatory arthritis by accelerating local lymphatic drainage in mice. Arthritis Rheum 63:2318–2328
86. Baldwin ME, Catimel B, Nice EC, Roufail S, Hall NE, Stenvers KL et al (2001) The specificity of receptor binding by vascular endothelial growth factor-d is different in mouse and man. J Biol Chem 276:19166–19171
87. Baldwin ME, Halford MM, Roufail S, Williams RA, Hibbs ML, Grail D et al (2005) Vascular endothelial growth factor D is dispensable for development of the lymphatic system. Mol Cell Biol 25:2441–2449
88. Rissanen TT, Markkanen JE, Gruchala M, Heikura T, Puranen A, Kettunen MI et al (2003) VEGF-D is the strongest angiogenic and lymphangiogenic effector among VEGFs delivered into skeletal muscle via adenoviruses. Circ Res 92:1098–1106
89. Byzova TV, Goldman CK, Jankau J, Chen J, Cabrera G, Achen MG et al (2002) Adenovirus encoding vascular endothelial growth factor-D induces tissue-specific vascular patterns in vivo. Blood 99:4434–4442
90. Stacker SA, Caesar C, Baldwin ME, Thornton GE, Williams RA, Prevo R et al (2001) VEGF-D promotes the metastatic spread of tumor cells via the lymphatics. Nat Med 7:186–191
91. Dumont DJ, Jussila L, Taipale J, Lymboussaki A, Mustonen T, Pajusola K et al (1998) Cardiovascular failure in mouse embryos deficient in VEGF receptor-3. Science 282:946–949
92. Irrthum A, Karkkainen MJ, Devriendt K, Alitalo K, Vikkula M (2000) Congenital hereditary lymphedema caused by a mutation that inactivates VEGFR3 tyrosine kinase. Am J Hum Genet 67:7–17
93. Karkkainen MJ, Ferrell RE, Lawrence EC, Kimak MA, Levinson KL, McTigue MA et al (2000) Missense mutations interfere with VEGFR-3 signalling in primary lymphoedema. Nat Genet 25:153–159
94. Partanen TA, Vuola P, Jauhiainen S, Lohi J, Salminen P, Pitkäranta A et al (2013) Neuropilin-2 and vascular endothelial growth factor receptor-3 are up-regulated in human vascular malformations. Angiogenesis 16(1):137–46
95. Martínez-Corral I, Olmeda D, Diéguez-Hurtado R, Tammela T, Alitalo K, Ortega S (2012) In vivo imaging of lymphatic vessels in development, wound healing, inflammation, and tumor metastasis. Proc Natl Acad Sci USA 109:6223–6228
96. Flister MJ, Wilber A, Hall KL, Iwata C, Miyazono K, Nisato RE et al (2010) Inflammation induces lymphangiogenesis through up-regulation of VEGFR-3 mediated by NF-kappaB and Prox1. Blood 115:418–429
97. Maruyama K, Ii M, Cursiefen C, Jackson DG, Keino H, Tomita M et al (2005) Inflammation-induced lymphangiogenesis in the cornea arises from CD11b-positive macrophages. J Clin Invest 115:2363–2372
98. Wigle JT, Oliver G (1999) Prox1 function is required for the development of the murine lymphatic system. Cell 98:769–778

99. Albrecht I, Christofori G (2011) Molecular mechanisms of lymphangiogenesis in development and cancer. Int J Dev Biol 55:483–494
100. Mäkinen T, Adams RH, Bailey J, Lu Q, Ziemiecki A, Alitalo K et al (2005) PDZ interaction site in ephrinB2 is required for the remodeling of lymphatic vasculature. Genes Dev 19:397–410
101. Niessen K, Zhang G, Ridgway JB, Chen H, Kolumam G, Siebel CW et al (2011) The Notch1-Dll4 signaling pathway regulates mouse postnatal lymphatic development. Blood 118:1989–1997
102. Zheng W, Tammela T, Yamamoto M, Anisimov A, Holopainen T, Kaijalainen S et al (2011) Notch restricts lymphatic vessel sprouting induced by vascular endothelial growth factor. Blood 118:1154–1162
103. Partanen TA, Mäkinen T, Arola J, Suda T, Weich HA, Alitalo K (1999) Endothelial growth factor receptors in human fetal heart. Circulation 100:583–586
104. Dumont DJ, Jussila L, Taipale J, Lymboussaki A, Mustonen T, Pajusola K et al (1998) Cardiovascular failure in mouse embryos deficient in VEGF receptor-3. Science 282:946–949
105. Geissler HJ, Bloch W, Förster SF, Mehlhorn U, Krahwinkel A, Kroener A et al (2003) Morphology and density of initial lymphatics in human myocardium determined by immunohistochemistry. Thorac Cardiovasc Surg 51:244–248
106. Geissler HJ, Dashkevich A, Fischer UM, Fries JWU, Kuhn-Régnier F, Addicks K et al (2006) First year changes of myocardial lymphatic endothelial markers in heart transplant recipients. Eur J Cardiothorac Surg 29:767–771
107. Lakkis FG, Arakelov A, Konieczny BT, Inoue Y (2000) Immunologic "ignorance" of vascularized organ transplants in the absence of secondary lymphoid tissue. Nat Med 6:686–688
108. Larsen CP, Morris PJ, Austyn JM (1990) Migration of dendritic leukocytes from cardiac allografts into host spleens. A novel pathway for initiation of rejection. J Exp Med 171:307–314
109. Lodge-Patch I (1951) The ageing of cardiac infarcts, and its influence on cardiac rupture. Br Heart J 13:37–42
110. Ishikawa Y, Akasaka Y, Ishii T, Itoh K, Masuda T, Zhang L et al (2000) Sequential changes in localization of repair-related proteins (heat shock protein 70, ubiquitin and vascular endothelial growth factor) in the different stages of myocardial infarction. Histopathology 37:546–554
111. Ishikawa Y, Komiyama K, Masuda S, Murakami M, Akasaka Y, Ito K et al (2005) Expression of type V secretory phospholipase A in myocardial remodelling after infarction. Histopathology 47:257–267
112. Banai S, Jaklitsch MT, Shou M, Lazarous DF, Scheinowitz M, Biro S, Epstein SE et al (1994) Angiogenic-induced enhancement of collateral blood flow to ischemic myocardium by vascular endothelial growth factor in dogs. Circulation 89:2183–2189
113. Ishikawa Y, Akishima-Fukasawa Y, Ito K, Akasaka Y, Tanaka M, Shimokawa R et al (2007) Lymphangiogenesis in myocardial remodelling after infarction. Histopathology 51:345–353
114. Libby P, Ridker PM, Hansson GK, Leducq Transatlantic Network on Atherothrombosis (2009) Inflammation in atherosclerosis: from pathophysiology to practice. J Am Coll Cardiol 54:2129–2138
115. Kholová I, Dragneva G, Cermáková P, Laidinen S, Kaskenpää N, Hazes T et al (2011) Lymphatic vasculature is increased in heart valves, ischaemic and inflamed hearts and in cholesterol-rich and calcified atherosclerotic lesions. Eur J Clin Invest 41:487–497
116. Gräbner R, Lötzer K, Döpping S, Hildner M, Radke D, Beer M et al (2009) Lymphotoxin beta receptor signaling promotes tertiary lymphoid organogenesis in the aorta adventitia of aged ApoE-/- mice. J Exp Med 206:233–248
117. Nakano T, Nakashima Y, Yonemitsu Y, Sumiyoshi S, Chen Y-X, Akishima Y et al (2005) Angiogenesis and lymphangiogenesis and expression of lymphangiogenic factors in the atherosclerotic intima of human coronary arteries. Hum Pathol 36:330–340

118. Mohler ER, Gannon F, Reynolds C, Zimmerman R, Keane MG, Kaplan FS (2001) Bone formation and inflammation in cardiac valves. Circulation 103:1522–1528
119. Otto CM, Kuusisto J, Reichenbach DD, Gown AM, O'Brien KD (1994) Characterization of the early lesion of "degenerative" valvular aortic stenosis: histological and immunohistochemical studies. Circulation 90:844–853
120. Soini Y, Salo T, Satta J (2003) Angiogenesis is involved in the pathogenesis of nonrheumatic aortic valve stenosis. Hum Pathol 34:756–763
121. Syväranta S, Helske S, Laine M, Lappalainen J, Kupari M, Mäyränpää MI et al (2010) Vascular endothelial growth factor-secreting mast cells and myofibroblasts: a novel self-perpetuating angiogenic pathway in aortic valve stenosis. Arterioscler Thromb Vasc Biol 30:1220–1227
122. Syväranta S, Helske S, Lappalainen J, Kupari M, Kovanen PT (2012) Lymphangiogenesis in aortic valve stenosis—novel regulatory roles for valvular myofibroblasts and mast cells. Atherosclerosis 221:366–374
123. Aharinejad S, Schäfer R, Hofbauer R, Abraham D, Blumer R, Miksovsky A et al (2001) Impact of cardiac transplantation on molecular pathology of ET-1, VEGF-C, and mitochondrial metabolism and morphology in dilated versus ischemic cardiomyopathic patients. Transplantation 72:1043–1049
124. Dashkevich A, Bloch W, Antonyan A, Goebel H, Fries JUW, Schlensak C et al (2010) Immunohistochemical study of remodeling of myocardial lymphatic and blood microvascular structures in terminal heart failure: differences between ischemic and dilated cardiomyopathy. Lymphology 43:110–117
125. Whitehurst B, Flister MJ, Bagaitkar J, Volk L, Bivens CM, Pickett B et al (2007) Anti-VEGF-A therapy reduces lymphatic vessel density and expression of VEGFR-3 in an orthotopic breast tumor model. Int J Cancer 121:2181–2191
126. Paavonen K, Puolakkainen P, Jussila L, Jahkola T, Alitalo K (2010) Vascular endothelial growth factor receptor-3 in lymphangiogenesis in wound healing. Am J Pathol 156:1499–1504
127. Kunstfeld R, Hirakawa S, Hong Y-K, Schacht V, Lange-Asschenfeldt B, Velasco P et al (2004) Induction of cutaneous delayed-type hypersensitivity reactions in VEGF-A transgenic mice results in chronic skin inflammation associated with persistent lymphatic hyperplasia. Blood 104:1048–1057
128. Cursiefen C, Chen L, Borges LP, Jackson D, Cao J, Radziejewski C et al (2004) VEGF-A stimulates lymphangiogenesis and hemangiogenesis in inflammatory neovascularization via macrophage recruitment. J Clin Invest 113:1040–1050
129. Iwata C, Kano MR, Komuro A, Oka M, Kiyono K, Johansson E et al (2007) Inhibition of cyclooxygenase-2 suppresses lymph node metastasis via reduction of lymphangiogenesis. Cancer Res 67:10181–10189
130. Ristimäki A, Narko K, Enholm B, Joukov V, Alitalo K (1998) Proinflammatory cytokines regulate expression of the lymphatic endothelial mitogen vascular endothelial growth factor-C. J Biol Chem 273:8413–8418
131. Johnson LA, Clasper S, Holt AP, Lalor PF, Baban D, Jackson DG (2006) An inflammation-induced mechanism for leukocyte transmigration across lymphatic vessel endothelium. J Exp Med 203:2763–2777
132. Saban MR, Mémet S, Jackson DG, Ash J, Roig AA, Israël A et al (2004) Visualization of lymphatic vessels through NF-kappaB activity. Blood 104:3228–3230
133. Veikkola T, Jussila L, Mäkinen T, Karpanen T, Jeltsch M, Petrova TV et al (2001) Signalling via vascular endothelial growth factor receptor-3 is sufficient for lymphangiogenesis in transgenic mice. EMBO J 20:1223–1231
134. Kang S, Lee S-P, Kim KE, Kim H-Z, Mémet S, Koh GY (2009) Toll-like receptor 4 in lymphatic endothelial cells contributes to LPS-induced lymphangiogenesis by chemotactic recruitment of macrophages. Blood 113:2605–2613

Chapter 10
Hypoxia and the Cardiac Lymphatic System

Todd R. Wuest and P. Charles Lin

Abstract Vertebrates are endowed with two separate and complimentary vascular systems. The blood vasculature serves to deliver nutrients and oxygen, while the lymphatic network acts to return extravasated fluid back to the blood circulation. The cardiac lymphatic network is at a unique intersection between these two systems. Impairment of the cardiac lymphatic drainage elicits severe myocardial edema and impairs blood distribution, highlighting the importance of this understudied area.

As the heart is at the center of the body's oxygen distribution system, cardiac tissue is intricately regulated by oxygen availability. The lymphatic network is affected by hypoxia too, which alters the expression of lymphangiogenic factors and promotes inflammatory responses. Under some circumstances, lymphatic responses to hypoxia are believed to beneficial to cardiac function through the relief of edema. However, lymphatic drainage influences the development of chronic and destructive inflammation, a root cause of many forms of cardiovascular diseases. In this chapter, we will review the mechanisms by which the lymphatic vasculature senses and responds to hypoxia with an emphasis on the cardiac lymphatic network.

Keywords Hypoxia • Lymphatic vessel • Lymphangiogenesis • Heart • VEGF

Introduction

The human vasculature faces a complex and fluctuating set of stresses that require continued growth, remodeling, and adaptation. Although the impact of ischemia and hypoxia on the cardiac blood vascular network has been extensively studied, historically, little attention has been paid to the ramifications of cardiac events on the

T.R. Wuest • P.C. Lin (✉)
Mouse Cancer Genetics Program, National Cancer Institute, 1050 Boyles Street, Bldg 560, Rm. 12-34, PO Box B, Frederick, MD 21702, USA
e-mail: p.lin@nih.gov

G. Karunamuni (ed.), *The Cardiac Lymphatic System: An Overview*,
DOI 10.1007/978-1-4614-6774-8_10,

lymphatic vasculature. Yet, the efficient function of the lymphatic system is required to prevent and resolve cardiac edema and preserve the sustained flow of oxygen and nutrients to heart tissue [1–3]. Coronary lymphatic blockade rapidly elicits myocardial edema, impairs contractile function, and contributes to cardiac failure, effects which emphasize the importance of the cardiac lymphatic system [4].

The study of the lymphatic network has accelerated as of late with the identification of specific cellular markers of the lymphatic vasculature such as LYVE-1, VEGFR-3, Prox1, and podoplanin as well as elucidation of many of the pathways regulating development lymphangiogenesis and function. These studies have shown broad homology between systems affecting the blood and lymphatic networks, including hypoxia. The cardiopulmonary system is the primary distributor of oxygen within the body and is regulated at the structural and functional level by oxygen availability [5, 6]. In the following chapter we will review the intersection between hypoxia and the lymphatic vasculature, specifically the underlying mechanisms of hypoxia's impact on lymphatic structure and function, as well as address unresolved questions within the field.

Biochemical Recognition of Hypoxia: The HIF System

The best-studied biochemical recognition pathway for hypoxia relies on a set of three distinct genetically encoded transcription factor heterodimers, coined as hypoxia-inducible factors 1, 2, and 3 [7]. The canonical member of the group, HIF-1, was first identified on the basis of its function as a cis-acting transcription factor regulating the activity of the 3′ erythropoietin enhancer [8, 9]. Shortly thereafter, it was found that HIF-1 in fact regulates a remarkable diversity of responses to hypoxia including the expression of angiogenic factors and inflammatory cytokines, as well as diverting cellular metabolic pathways from oxidative to glycolytic energy production [7, 9, 10].

The rapidity of physiological changes in oxygen availability necessitates a correspondingly rapid flux in the activity of oxygen-sensing systems. Hypoxic induction of cellular gene expression is detectable within minutes, and the body-wide expression of HIF-1 targets such as erythropoietin may change several orders of magnitude within the span of just a few hours [7, 8]. The rapid time course of hypoxia-driven gene transcription contrasts dramatically with systems regulating developmental angiogenesis and lymphangiogenesis [11–13]. The uniquely rapid impact of hypoxia emphasizes its potential importance on the cardiac lymphatic system.

HIFs are composed of sets of paired α and β subunits, with unique pairs for each of the three HIFs [7, 14]. Of the three, HIF-1 and HIF-2 function to activate hypoxia target genes, while HIF-3 functions primarily as an isoform-specific repressor of HIF-1 activity [7, 14]. In all of the HIFs, the β subunit is constitutively present within the cell; however, the availability and activity of corresponding α subunits are subject to layers of regulatory control on the basis of cellular oxygen levels [7, 14]. HIFα subunits are constitutively transcribed and translated yet are rapidly degraded under normoxic conditions through the activity of a set of three proline hydroxylases (PHD1-3) [15]. Under normoxic conditions, PHDs catalyze the oxygen dependent hydroxylation of HIFα subunits [15]. Hydroxylated HIFα subunits are then susceptible to recognition by the von Hippel-Lindau (VHL) tumor suppressor protein at either of two conserved proline residues [7, 16]. The E3 ubiquitin ligase activity of the VHL complex then targets bound HIFα for rapid degradation, maintaining low cellular concentrations.

Hypoxia impairs the prolyl hydroxylation required for VHL recognition, thus reducing degradation and resulting in a rapid increase in cellular HIFα levels [7, 14]. Hypoxic repression of prolyl hydroxylation allows accumulation of HIFα subunits, which are then free to dimerize with corresponding HIFβ subunits [7, 14]. Dimerization of the two subunits promotes nuclear accumulation of the HIF complex and then activates transcription of promoters containing hypoxia response elements (A/G CGTG), which includes angiogenic and lymphangiogenic growth factors [17–19].

Questions remain regarding the exact mechanism by which hypoxia drives HIF nuclear accumulation. HIF-1α and β subunits are capable of nuclear translocation even in the absence of dimerization [20]. A recent study has also provided evidence that prolyl hydroxylase 2 (PHD2) primarily acts on nuclear HIF-1α [21]. It also appears that VHL exerts its repressive activity on HIF-1 function from within the nucleus as a VHL mutant strictly localized to the nucleus was capable of fully recapitulating wild-type VHL activity, while cytoplasmically localized mutants were devoid of such activity [22]. Both pieces of evidence suggest that substantial nuclear translocation of HIF-1α must occur during both hypoxic and normoxic conditions. Yet, HIF-1α only accumulates within the nucleus during hypoxia [23], suggesting that accumulation is regulated by altering the rate of nuclear export.

Not surprisingly, evolution has endowed the HIF system with multiple layers of regulatory control. Constitutive stabilization of HIF-1α and dimerization with HIF-1β is insufficient to drive the transcription of target genes in the absence of an additional signal provided by hypoxia [24]. The mechanism by which hypoxia relieves repression of HIF-1 is still a subject of active investigation, but the field was considerably advanced with the discovery of a novel factor inhibiting HIF-1 and HIF-2α via asparagine rather than prolyl hydroxylation. Rather than blocking HIFα nuclear accumulation, factor inhibiting HIF-1 (FIH-1) catalyzes the oxygen dependent hydroxylation of a conserved asparagine residue in the c-terminal transactivation domain of HIF-1 and HIF-2α [25]. Asparagine hydroxylation blocks the recruitment of CBP/p300 to the HIF complex and effectively silences transcription activation by HIF-1 and HIF-2 [26, 27] (Fig. 10.1).

Intriguingly, HIFα subunits are not the only target of FIH-1. FIH-1 also binds asparagine hydroxylates and inactivates Notch1–4 with even greater avidity than HIF-1α. In fact, the binding affinity of FIH-1 for Notch1 is roughly 250-fold higher than for HIF-1α [28]. Binding affinities also differ by a full order of magnitude for FIH-1 between individual members of the Notch family [28]. Notch signaling elegantly regulates endothelial sprouting by coordinating tip versus stalk cell fate determination along with lymphangiogenic growth factor responses [29, 30]. The differing binding constants for HIFα subunits and Notch for FIH-1 suggest a

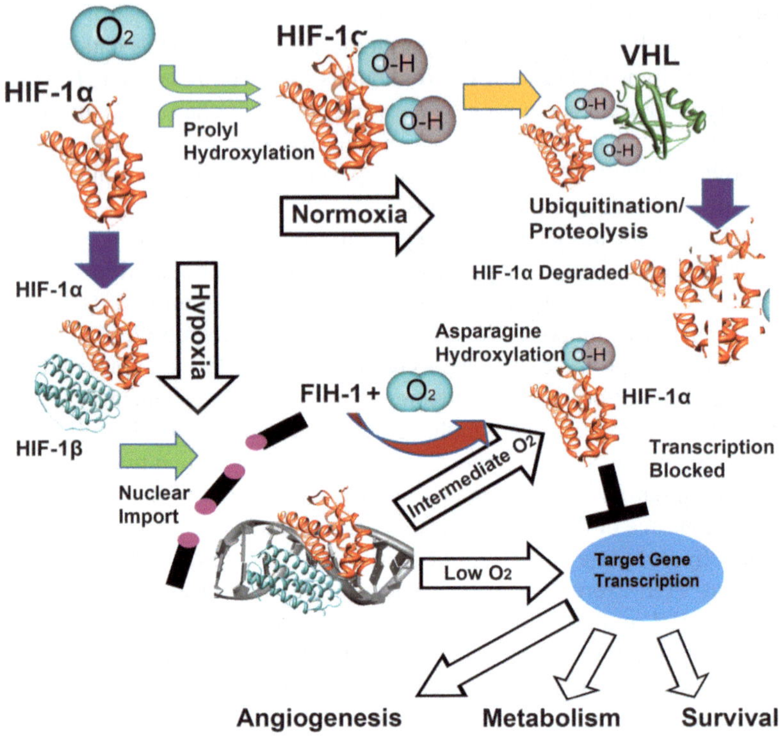

Fig. 10.1 Diagram of hypoxic regulation of HIF-1. Hypoxia regulates the activity of HIF-1 via two discrete pathways. Prolyl hydroxylases 1–3 catalyze the oxygen dependent hydroxylation of HIF-1α at two conserved proline residues. Prolyl hydroxylation allows recognition of HIF-1α by the VHL ubiquitin ligase complex. Ubiquitin ligation leads to proteolytic degradation and keeps HIF-1α levels low during normoxic conditions. Inhibition of prolyl hydroxylation due to hypoxia leads to the accumulation of HIF-1α, dimerization with HIF-1β, and nuclear accumulation. In the event of intermediate oxygen levels, FIH-1 catalyzes the asparagine hydroxylation of nuclear HIF-1α and blocks transcriptional activation by nuclear HIF-1. Only under conditions in which both prolyl hydroxylation and asparagine hydroxylation are inhibited by low oxygen concentrations is HIF-1 capable of activating transcription of target genes

mechanism by which FIH-1 may affect endothelial structure according to oxygen tension. Under normoxic conditions, PHD and FIH-1 concomitantly repress both HIF and Notch. But, as available oxygen decreases, repression of HIF by PHD and FIH-1 is relieved prior to repression of Notch owing to the lower activity of HIF/PHD/VHL and HIF/FIH-1 relative to Notch blockade by FIH-1 [7, 31]. Thus, hypoxia affects endothelial cell growth via HIF-mediated growth factor expression and independently influences endothelial organization and maturation via conditional repression of Notch (Fig. 10.2).

Currently, little is known about how HIF and Notch coordination may affect lymphatic growth and function during cardiac disease. But owing to the high degree of analogy between venous and lymphatic systems and the shared reliance on

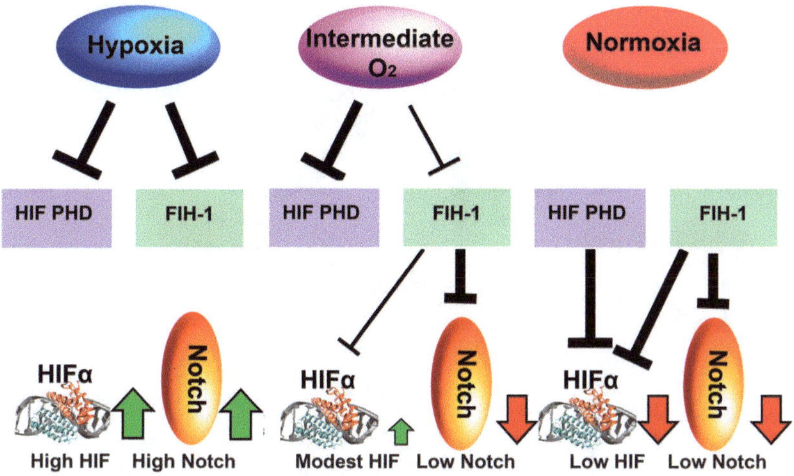

Fig. 10.2 Fine tuning of the hypoxic response. PHD and FIH-1 pathways differentially regulate the activity of HIFs and Notch. PHDs 1–3 require higher oxygen tensions than FIH-1 for activity. FIH-1 also has a higher affinity for Notch than HIF-1 and HIF-2α. These two factors are believed to fine tune the response to hypoxia. During hypoxia, repression of PHDs and FIH-1 allows activation of both HIF-1/2 and Notch. But intermediate oxygen tensions preferentially block the activity of Notch, while still allowing HIF activity

HIF-regulated growth factors, such as vascular endothelial growth factors (VEGFs) and its receptors, between the two systems, it is logical to assume a correlation. HIF-1 mediates cardiac protection during ischemia and increases vascular bed development following infarction [32, 33]. Presumably, lymphatic vessel growth is also promoted by HIF-1 activation, but it remains to be seen how significant the lymphatic impact of HIF-1 is on the development of cardiac edema.

Lymphangiogenic Growth Factors and Hypoxia

The VEGF Family

A variety of hypoxia target genes elicit the growth and maturation of lymphatic vessels, but among the most clearly defined are members of the VEGF family. The founding member of the group, VEGF-A, was first identified in the laboratory of Harold Dvorak as a tumor-secreted factor capable of inducing increased vascular permeability. Appropriately, the protein was initially coined as vascular permeability factor (VPF) [34]. It was not until a seminal study by Napoleon Ferrara's group that VEGF-A was identified as a potent and specific endothelial mitogen [35].

The human genome encodes five VEGFs: VEGF-A, VEGF-B, VEGF-C, VEGF-D, and placental growth factor (PGF). Alternative mRNA splicing and

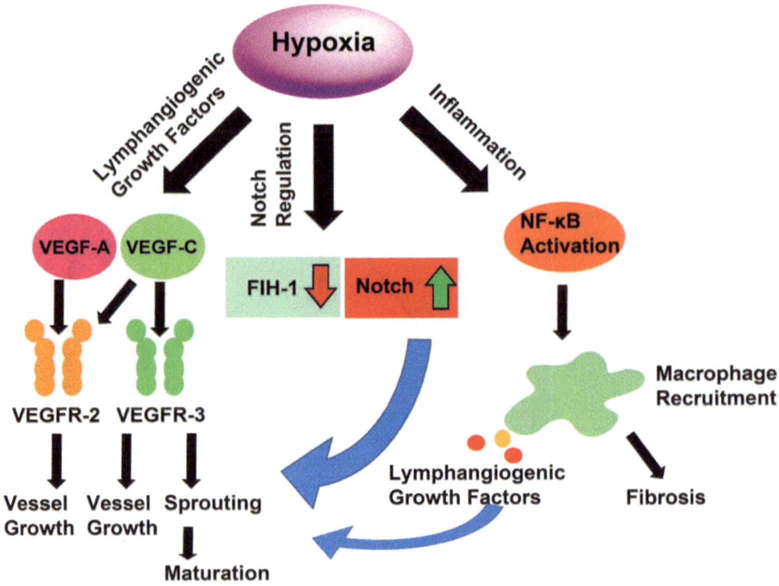

Fig. 10.3 Overview of hypoxia and lymphangiogenesis. Hypoxia affects the lymphatic vasculature at multiple levels. The activation of HIF-1 and HIF-2 as well as NF-κB regulates the expression of lymphangiogenic growth factors. Hypoxic relief of Notch repression also allows for increased Notch signaling

proteolytic processing considerably expand the numbers of individual factors, and the human VEGF-A gene alone codes for seven distinct proangiogenic and five antiangiogenic isoforms [36]. The VEGFs exert their functions through three related receptor tyrosine kinases: VEGFR-1 (binds VEGF-A, VEGF-B, and PGF), VEGFR-2 (binds VEGF-A and proteolytically processed forms of VEGF-C and VEGF-D), and VEGFR-3 (binds VEGF-C and VEGF-D) as well as two co-receptors neuropilin-1 and neuropilin-2 [37]. The binding of ligand to receptor induces dimerization and receptor kinase activation which in turn phosphorylates the receptor and downstream signal transducing molecules [37]. Signal transduction is further modulated by co-receptor interactions with neuropilins [37–39].

As with VEGF-A, the other VEGFs are powerful mitogens but their effective target cell range differs according to cellular VEGF receptor expression [37]. Of the three VEGF receptors, adult lymphatic vessels normally express only VEGFR-2 and VEGFR-3 and the co-receptor neuropilin-2 [11–13]. Treatment of lymphatic vessels with the specific VEGFR-1 agonist PGF does not produce either lymphatic hypoplasia or sprouting [40]. Thus, of the VEGF family, only the VEGFR-2 and VEGFR-3 ligands: VEGF-A, VEGF-C, and VEGF-D, directly stimulate lymphangiogenesis (Fig. 10.3). However, VEGFR-1 ligands indirectly affect lymphatic vessel structure through the recruitment of VEGF-C/D expressing inflammatory leukocytes [41].

There is considerable controversy regarding the relative contribution between VEGF-A, VEGF-C, and VEGF-D to lymphatic growth and maturation. Lymphatic vascular development and maintenance strictly requires VEGFR-3/VEGF-C, and the absence of either leads to embryonic lethality due to lymphatic insufficiency [12, 13]. VEGF-D is also a potent lymphatic growth factor upregulated under hypoxic conditions [42]. But the impact of VEGF-A acting through VEGFR-2 is subject to debate. It is clear that VEGF-A overexpression leads to lymphangiogenesis [43, 44]. In some settings, VEGF-A-driven lymphangiogenesis appears to be the indirect result of recruitment of inflammatory macrophages [45]. Inflammatory conditions drive macrophage expression of VEGF-C/D which then stimulates lymphatic growth through VEGFR-3 activation [45]. This indirect model for VEGF-A/VEGFR-2-mediated lymphangiogenesis, in which VEGF-A acts primarily as a chemoattractant, is supported by the observation that the treatment with parapoxvirus-encoded VEGFR-2-specific agonist, VEGF-E, causes only lymphatic dilation without sprouting [40].

However, VEGF-A/VEGFR-2 is capable of eliciting lymphatic vessel growth and sprouting even in the absence of VEGF-C expression and monocyte recruitment [44]. Even though VEGF-E promotes lymphatic dilation without sprouting, this growth factor interacts with VEGFR-2 and only the VEGF co-receptor neuropilin-1, not neuropilin-2 [46]. Lymphatic vessels typically express only neuropilin-2, and neuropilin-2 is required for lymphatic sprouting and maturation [47]. VEGF-A is capable of neuropilin-2 interaction [46]. Therefore, it is reasonable to suspect that the absence of lymphatic sprouts following VEGF-E is not directly relevant to the in vivo activities of VEGF-A. As VEGF-A is capable of driving lymphangiogenesis in vitro and in vivo, and VEGF-A expression is strongly upregulated by hypoxia, it is logical to assume that VEGF-A makes direct and indirect contributions to the growth and structure of the lymphatic system during hypoxic conditions. In fact, VEGF-A-driven lymphangiogenesis may be of particular importance in the cardiac system. The cardiac lymphatic system is reported to uniquely express a population of vessels with lymphatic morphology and no detectable VEGFR-3 expression [48].

Lymphangiogenic Growth Factor Expression During Hypoxia

Hypoxia elicits the expression of all human VEGFs [42, 49, 50]. However, the degree, duration, and cell-specific expression differ among the VEGFs, suggesting that the mechanism by which hypoxia directs expression also differs. In the case of VEGF-A, hypoxia simultaneously affects promoter activity, transcript stability, and translation efficiency [51]. The increase in VEGF-A promoter activity is primarily a consequence of hypoxic activation of the HIF system [52]. VEGF-A was among the first genes for which a HIF-1-binding element was discovered, and deletion of the HRE within the VEGF-A promoters blocks transcriptional upregulation during hypoxia [53]. A hypoxia response element has also been identified in the promoters of the lymphangiogenic growth factors, platelet-derived growth factor (PDGF-B), and fibroblast growth factor (FGF)-2 [54, 55].

Despite transcriptional upregulation of VEGF-C and VEGF-D by hypoxia, neither human gene contains a HRE suggesting that ancillary hypoxia dependent pathways are responsible [56, 57]. Both genes are subject to upregulation following NF-κB activation [58]. The promoters for VEGF-C and VEGF-D both contain numerous NF-κB binding sites, and a blockade of NF-κB function abrogates expression of VEGF-C/D during chronic inflammation and reduces lymphangiogenesis [58]. Hypoxia activates NF-κB, though reports differ slightly regarding the mechanism by which this occurs. In neutrophils, hypoxic activation of HIF-1 promotes the transcription of NF-κB p65 and the NF-κB activator, IKKα [59]. A recent study from the lab of Sonia Rocha has also implicated the IKK complex in hypoxic activation of NF-κB but provided evidence that calcium release from endoplasmic reticulum stores is a critical initiator via activation of CamK2 and subsequent activation of the ubiquitin dependent IKK kinase, TAK1 [60]. In addition to promoting VEGF-C and VEGF-D expression, NF-κB activation also upregulates the expression of VEGFR-2 and VEGFR-3 which increases lymphatic susceptibility to exogenous growth factor [61, 62].

Hypoxia also activates members of the CCAAT-enhancer-binding protein (C/EBP) family of transcription factors. C/EBPβ and δ both cooperatively interact with NF-κβ constituents to activate transcription [63, 64]. The human VEGF-C promoter contains multiple CCAAT sites [65], and C/EBPδ functions to promote expression of both VEGF-C and its receptor VEGFR-3 during hypoxia in a HIF-1-dependent fashion [66]. Hypoxic induction of C/EBPβ also acts to increase production of iNOS [67]. The expression of iNOS correlates with increased contractile activity of collecting lymphatics [68] and suggests an avenue by which hypoxia may affect not only the growth and structure of lymphatic vessels but also fluid transport capability via regulation of collecting vessel contraction.

It has been demonstrated that administration of lymphangiogenic growth factors improves the outcome following myocardial infarction. For example, a treatment with VEGF-A improves cardiac function and viability in canine and porcine infarction models [69–71]. Due to the mixed angiogenic and lymphangiogenic effects of VEGF-A, it is unclear how great the contribution of improved lymphatic function is to the protection of myocardial infarction. But, notably, administration of VEGF-C and VEGF-D also improves outcomes after cardiac infarction [72, 73]. VEGF-C and VEGF-D exert primarily on lymphangiogenic rather than angiogenic effects [40, 74] and the efficacy of these agents implicates lymphangiogenesis as a potential therapeutic process to resolve the complications of infarction.

Hypoxia, Inflammation, and the Deleterious Impact of Lymphatics

The concert of hypoxia-induced production of growth factors during infarction provides critical protection to cardiac tissue. Endogenously expressed lymphangiogenic growth factors serve to promote damage resolution and exogenous administration may provide potential therapeutic avenues to improve cardiac function under such circumstances. However, during chronic inflammation, hypoxia and lymphangiogenesis are believed to exacerbate the inflammatory response that may impair cardiac function [75–77].

Chronic inflammation has long been implicated in the pathogenesis of heart disease. The activation of NF-κB by hypoxia leads to cardiovascular overexpression of NF-κB-dependent cytokines [78]. Elevated levels of inflammatory markers such as IL-1, TNFα, and C-reactive protein are independently associated with poor prognosis [79]. Leukocyte infiltration into cardiovascular tissue is a common phenomenon during cardiovascular disease, and abrogation of leukocyte recruitment improves experimental measures of cardiovascular function [79, 80].

Lymphangiogenesis has been implicated in the development of chronic inflammation in other areas of the body, albeit indirectly. One of the hallmarks of immunologically privileged tissues is the absence of lymphatic vessels [81, 82]. Conversely, lymphatic vessel development in previously immunologically privileged tissues, such as the cornea, is associated with the loss of privilege and a chronically inflamed state [83]. Removal of lymphatic drainage from the cornea by excision of draining lymph nodes blocks corneal allograft rejection, strongly implicating a requirement for lymphatic drainage in the development of chronic inflammatory responses [84]. Lymphangiogenesis also correlates with organ rejection following cardiac transplant, and the administration of anti-lymphangiogenic treatments impairs rejection and ameliorates arteriosclerosis [85, 86]. This suggests that, as is the case with other organs, lymphangiogenic factor expression during inflammatory cardiac disease could aggravate the inflammatory process.

Although the heart is well drained by lymphatic vessels, the valve tissue is normally sparsely drained by the lymphatic system [86–88]. Valvular stenosis is associated with increased expression of VEGF-A, VEGF-C, and VEGF-D and increased lymphatic vessel density [86]. Adventitial inflammation via balloon-induced injury has also been shown to upregulate the expression of lymphangiogenic factors, and increased adventitial lymphatic vessel density correlates with intimal hyperplasia [89]. Taken together, the linkage between lymphangiogenesis and inflammatory cardiovascular illness imply that hypoxia-driven lymphangiogenesis may have two very different faces in the heart. While lymphangiogenesis helps to resolve myocardial edema during acute hypoxic insult, chronic hypoxia and associated inflammation may synergize with increased lymphatic drainage to prolong and aggravate heart-damaging inflammatory processes [78–80, 90].

Conclusions and Future Directions

The relative paucity of data specific to the cardiac lymphatic system leaves an abundance of unresolved questions. First and foremost is the question of how lymphatic vessel growth effects long-term cardiovascular disease. Even though hypoxia-driven lymphangiogenesis appears beneficial during acute hypoxia, acute hypoxic events in human patients are usually the pinnacle of long-term cardiovascular disease, a situation in which lymphangiogenesis may be deleterious. It is difficult to adequately separate variables in the fashion required to study such a situation. For example, dendritic cells variably express all three VEGF receptors [91–93]. Administration of VEGF-A suppresses dendritic cell function [94]. Conversely, administration of VEGF-C is associated with dendritic cell activation and lymphatic egress [95]. Thus, the cross-talk between the lymphatic and immune systems precludes experimentally altering one system without affecting the other.

The nature and impact of the HIF-1 system on resolution of cardiac hypoxia is also subject to debate. Although hypoxia activates HIF-1 in a broad range of cardiac cell types [90], the relative contribution of each cell to responses activated by HIF-1 is unclear. Genetic deletion of HIF-1α results in embryonic lethality, complicating experimental analysis of the HIF-1 system in vivo [96]. Some progress has been made recently through the use of conditional cre-lox-mediated deletion. Endothelial-specific deletion of HIF-1 via cre expression under the Tie2 promoter leads to impaired cardiac function following transaortic constriction [97]. Hopefully, future studies will expand to decipher the specific role of HIF-1 in lymphatic endothelial cells via cre-mediated deletion under a lymphatic specific promoter such as Prox1 or podoplanin.

Another unresolved question is how hypoxia may affect the structure and function of lymphatic vessels, in and outside of the heart. Lymphatic vessels are organized into a set of three sequential phenotypes, with interstitial fluid first entering via lymphatic capillaries, pooling to pre-collecting lymphatic vessels, and then finally accumulating into large, valved collecting lymphatic vessels [98]. Efficient lymphatic drainage requires not only that interstitial fluid is drained by lymphatic capillaries but that the collected fluid is unidirectionally transported in collecting lymphatic vessels. Notch signaling regulates lymphatic sprouting and collecting vessel maturation [99, 100]. The obvious question is whether hypoxia, through its impact on Notch signal transduction, affects lymphatic vessel structure. Extrapolation from in vivo studies on blood vasculature would suggest that, yes, hypoxia impacts endothelial structure through notch-mediated regulation of sprouting, but this has not been tested specifically in lymphatic vessels in vivo [101]. However, in vitro studies of isolated lymphatic endothelial cells indicate that hypoxic regulation of notch at least modulates lymphatic endothelial VEGFR-2 expression [102]. Conceivably, hypoxia may also affect lymphatic valve structure. The transcription factor FoxC2 is required for lymphatic valve formation and maintenance [103]. Even heterozygous genetic insufficiency at the FoxC2 locus impairs lymphatic drainage [103]. FoxC2 is induced by hypoxia [104], begging the question of whether hypoxia may strengthen lymphatic valves as a result of FoxC2-mediated upregulation of valve leaflet genes.

Ultimately, the purpose of biological inquiry is not just the satisfaction of intellectual curiosity but also the development of therapies to cure human diseases. Doubtlessly, the scientific, medical, and patient communities are eagerly awaiting new therapeutic avenues to treat heart diseases. Current literature suggests that targeting of the hypoxia/lymphatic axis may ameliorate disease. However, such treatments come with the caution that short-term improvements might provoke long-term consequences if hypoxia and associated lymphangiogenesis lead to a promotion of chronic inflammation.

References

1. Loukas M, Abel N, Tubbs RS, Grabska J, Birungi J, Anderson RH (2011) The cardiac lymphatic system. Clin Anat 24:684–691
2. Cui Y (2010) Impact of lymphatic vessels on the heart. Thorac Cardiovasc Surg 58(1):1–7
3. Nakamura K, Rockson SG (2008) The role of the lymphatic circulation in the natural history and expression of cardiovascular disease. Int J Cardiol 129(3):309–317
4. Solti F, Lengyel E, Jellinek H, Schneider F, Juhász-Nagy A, Kékesi V (1994) Coronary arteriopathy after lymphatic blockade: an experimental study in dogs. Lymphology 27(4):173–180
5. Bernhardt WM, Warnecke C, Willam C, Tanaka T, Wiesener MS, Eckardt KU (2007) Organ protection by hypoxia and hypoxia-inducible factors. Methods Enzymol 435:221–245
6. Giaccia A, Siim BG, Johnson RS (2003) HIF-1 as a target for drug development. Nat Rev Drug Discov 2(10):803–811
7. Greer SN, Metcalf JL, Wang Y, Ohh M (2012) The updated biology of hypoxia-inducible factor. EMBO J 31(11):2448–2460
8. Semenza GL, Wang GL (1992) A nuclear factor induced by hypoxia via de novo protein synthesis binds to the human erythropoietin gene enhancer at a site required for transcriptional activation. Mol Cell Biol 12(12):5447–5454
9. Wang GL, Semenza GL (1993) General involvement of hypoxia-inducible factor 1 in transcriptional response to hypoxia. PNAS 90(9):4304–4308
10. Knighton DR, Hunt TK, Scheuenstuhl H, Halliday BJ, Werb Z, Banda MJ (1983) Oxygen tension regulates the expression of angiogenesis factor by macrophages. Science 221(4617):1283–1285
11. Ji RC (2008) Lymphatic endothelial cells, lymphedematous lymphangiogenesis, and molecular control of edema formation. Lymphat Res Biol 6(3–4):123–137
12. Oliver G, Srinivasan RS (2008) Lymphatic vasculature development: current concepts. Ann N Y Acad Sci 1131:75–81
13. Karpanen T, Alitalo K (2008) Molecular biology and pathology of lymphangiogenesis. Annu Rev Pathol 3:367–397
14. Pugh CW, Ratcliffe PJ (2003) Regulation of angiogenesis by hypoxia: role of the HIF system. Nat Med 9(6):677–684
15. Bruick RK, McKnight SL (2001) A conserved family of prolyl-4-hydroxylases that modify HIF. Science 294(5545):1337–1340
16. Ivan M, Kondo K, Yang H, Kim W, Valiando J, Ohh M et al (2001) HIFalpha targeted for VHL-mediated destruction by proline hydroxylation: implications for O2 sensing. Science 292(5516):464–468
17. Wenger RH, Gassmann M (1997) Oxygen(es) and the hypoxia-inducible factor-1. Biol Chem 7:609–616
18. Zampell JC, Yan A, Avraham T, Daluvoy S, Weitman ES, Mehrara BJ (2012) HIF-1α coordinates lymphangiogenesis during wound healing and in response to inflammation. FASEB J 3:1027–1039
19. Liu Y, Cox SR, Morita T, Kourembanas S (1995) Hypoxia regulates vascular endothelial growth factor gene expression in endothelial cells. Identification of a 5′ enhancer. Circ Res 77(3):638–643
20. Chilov D, Camenisch G, Kvietikova I, Ziegler U, Gassmann M, Wenger RH (1999) Induction and nuclear translocation of hypoxia-inducible factor-1 (HIF-1): heterodimerization with ARNT is not necessary for nuclear accumulation of HIF-1alpha. J Cell Sci Pt 8:1203–1212
21. Pientka FK, Hu J, Schindler SG, Brix B, Thiel A, Joehren O et al (2012) Oxygen sensing by Prolyl-4-Hydroxylase PHD2 within the nuclear compartment and the influence of compartimentalisation on HIF-1 signalling. J Cell Sci 125:5168–5176
22. Lewis MD, Roberts BJ (2003) Role of nuclear and cytoplasmic localization in the tumour-suppressor activity of the von Hippel-Lindau protein. Oncogene 22:3992–3997

23. Chilov D, Camenisch G, Kvietikova I, Ziegler U, Gassmann M, Wenger RH (1999) Induction and nuclear translocation of hypoxia-inducible factor-1 (HIF-1): heterodimerization with ARNT is not necessary for nuclear accumulation of HIF-1alpha. J Cell Sci 112(Pt 8):1203–1212
24. Tanimoto K, Makino Y, Pereira T, Poellinger L (2000) Mechanism of regulation of the hypoxia-inducible factor-1 alpha by the von Hippel-Lindau tumor suppressor protein. EMBO J 19(16):4298–4309
25. Lando D, Peet DJ, Whelan DA, Gorman JJ, Whitelaw ML (2002) Asparagine hydroxylation of the HIF transactivation domain: a hypoxic switch. Science 295:858–861
26. Lando D, Peet DJ, Gorman JJ, Whelan DA, Whitelaw ML, Bruick RK (2002) FIH-1 is an asparaginyl hydroxylase enzyme that regulates the transcriptional activity of hypoxia-inducible factor. Genes Dev 16:1466
27. Hewitson KS, McNeill LA, Elkins JM, Schofield CJ (2003) The role of iron and 2-oxoglutarate oxygenases in signalling. Biochem Soc Trans 31:510
28. Wilkins SE, Hyvärinen J, Chicher J, Gorman JJ, Peet DJ, Bilton RL et al (2009) Differences in hydroxylation and binding of Notch and HIF-1alpha demonstrate substrate selectivity for factor inhibiting HIF-1 (FIH-1). Int J Biochem Cell Biol 41(7):1563–1571
29. Niessen K, Zhang G, Ridgway JB, Chen H, Kolumam G, Siebel CW et al (2011) The Notch1-Dll4 signaling pathway regulates mouse postnatal lymphatic development. Blood 118(7):1989–1997
30. Zheng W, Tammela T, Yamamoto M, Anisimov A, Holopainen T, Kaijalainen S (2011) Notch restricts lymphatic vessel sprouting induced by vascular endothelial growth factor. Blood 118(4):1154–1162
31. Johnson EA (2011) HIF takes it up a notch. Sci Signal 4(181):33
32. Ong SG, Hausenloy DJ (2012) Hypoxia-inducible factor as a therapeutic target for cardioprotection. Pharmacol Ther 136(1):69–81
33. Kido M, Du L, Sullivan CC, Li X, Deutsch R, Jamieson SW, Thistlethwaite PA (2005) Hypoxia-inducible factor 1-alpha reduces infarction and attenuates progression of cardiac dysfunction after myocardial infarction in the mouse. J Am Coll Cardiol 46(11):2116–2124
34. Senger DR, Galli SJ, Dvorak AM, Perruzzi CA, Harvey VS, Dvorak HF (1983) Tumor cells secrete a vascular permeability factor that promotes accumulation of ascites fluid. Science 219(4587):983–985
35. Leung DW, Cachianes G, Kuang WJ, Goeddel DV, Ferrara N (1989) Vascular endothelial growth factor is a secreted angiogenic mitogen. Science 246(4935):1306–1309
36. Woolard J, Bevan HS, Harper SJ, Bates DO (2009) Molecular diversity of VEGF-A as a regulator of its biological activity. Microcirculation 16(7):572–592
37. Matsumoto T, Claesson-Welsh L (2001) VEGF receptor signal transduction. Sci STKE 112:21
38. Soker S, Takashima S, Miao HQ, Neufeld G, Klagsbrun M (1998) Neuropilin-1 is expressed by endothelial and tumor cells as an isoform-specific receptor for vascular endothelial growth factor. Cell 92(6):735–745
39. Uniewicz KA, Fernig DG (2008) Neuropilins: a versatile partner of extracellular molecules that regulate development and disease. Front Biosci 13:4339–4360
40. Wirzenius M, Tammela T, Uutela M, He Y, Odorisio T, Zambruno G, Nagy JA et al (2007) Distinct vascular endothelial growth factor signals for lymphatic vessel enlargement and sprouting. J Exp Med 204(6):1431–1440
41. Murakami M, Zheng Y, Hirashima M, Suda T, Morita Y, Ooehara J et al (2008) VEGFR1 tyrosine kinase signaling promotes lymphangiogenesis as well as angiogenesis indirectly via macrophage recruitment. Arterioscler Thromb Vasc Biol 28(4):658–664
42. Ikeda Y, Yonemitsu Y, Onimaru M, Nakano T, Miyazaki M, Kohno R et al (2006) The regulation of vascular endothelial growth factors (VEGF-A, -C, and -D) expression in the retinal pigment epithelium. Exp Eye Res 83(5):1031–1040
43. Nagy JA, Vasile E, Feng D, Sundberg C, Brown LF, Detmar MJ et al (2002) Vascular permeability factor/vascular endothelial growth factor induces lymphangiogenesis as well as angiogenesis. J Exp Med 196(11):1497–1506

44. Wuest TR, Carr DJ (2010) VEGF-A expression by HSV-1-infected cells drives corneal lymphangiogenesis. J Exp Med 207(1):101–115
45. Cursiefen C, Chen L, Borges LP, Jackson D, Cao J, Radziejewski C et al (2004) VEGF-A stimulates lymphangiogenesis and hemangiogenesis in inflammatory neovascularization via macrophage recruitment. J Clin Invest 113(7):1040–1050
46. Narazaki M, Segarra M, Tosato G (2008) Neuropilin-2: a new molecular target for antiangiogenic and antitumor strategies. J Natl Cancer Inst 100(2):81–83
47. Xu Y, Yuan L, Mak J, Pardanaud L, Caunt M, Kasman I et al (2010) Neuropilin-2 mediates VEGF-C-induced lymphatic sprouting together with VEGFR3. J Cell Biol 188(1):115–130
48. Dashkevich A, Bloch W, Antonyan A, Fries JU, Geissler HJ (2009) Morphological and quantitative changes of the initial myocardial lymphatics in terminal heart failure. Lymphat Res Biol 7(1):21–27
49. Sands M, Howell K, Costello CM, McLoughlin P (2011) Placenta growth factor and vascular endothelial growth factor B expression in the hypoxic lung. Respir Res 25:12–17
50. Enholm B, Paavonen K, Ristimäki A, Kumar V, Gunji Y, Klefstrom J et al (1997) Comparison of VEGF, VEGF-B, VEGF-C and Ang-1 mRNA regulation by serum, growth factors, oncoproteins and hypoxia. Oncogene 14(20):2475–2483
51. Jafarifar F, Yao P, Eswarappa SM, Fox PL (2011) Repression of VEGFA by CA-rich element-binding microRNAs is modulated by hnRNP L. EMBO J 30(7):1324–1334
52. Tsuzuki Y, Fukumura D, Oosthuyse B, Koike C, Carmeliet P, Jain RK (2007) Vascular endothelial growth factor (VEGF) modulation by targeting hypoxia-inducible factor-1alpha–> hypoxia response element–> VEGF cascade differentially regulates vascular response and growth rate in tumors. Cancer Res 60(22):6248–6252
53. Pagès G, Pouysségur J (2005) Transcriptional regulation of the Vascular Endothelial Growth Factor gene—a concert of activating factors. Cardiovasc Res 65(3):564–573
54. Ullerås E, Wilcock A, Miller SJ, Franklin GC (2011) The sequential activation and repression of the human PDGF-B gene during chronic hypoxia reveals antagonistic roles for the depletion of oxygen and glucose. Growth Factors 19(4):233–245
55. Black SM, DeVol JM, Wedgwood S (2008) Regulation of fibroblast growth factor-2 expression in pulmonary arterial smooth muscle cells involves increased reactive oxygen species generation. Am J Physiol Cell Physiol 294(1):C345–C354
56. Yonekura H, Sakurai S, Liu X, Migita H, Wang H, Yamagishi S, Nomura M et al (2009) Placenta growth factor and vascular endothelial growth factor B and C expression in microvascular endothelial cells and pericytes. Implication in autocrine and paracrine regulation of angiogenesis. J Biol Chem 274(49):35172–35178
57. Ross MT, Grafham DV, Coffey AJ, Scherer S, McLay K, Muzny D et al (2005) The DNA sequence of the human X chromosome. Nature 434(7031):325–337
58. Watari K, Nakao S, Fotovati A, Basaki Y, Hosoi F, Bereczky B et al (2008) Role of macrophages in inflammatory lymphangiogenesis: Enhanced production of vascular endothelial growth factor C and D through NF-kappaB activation. Biochem Biophys Res Commun 377(3):826–831
59. Walmsley SR, Print C, Farahi N, Peyssonnaux C, Johnson RS, Cramer T et al (2005) Hypoxia-induced neutrophil survival is mediated by HIF-1alpha-dependent NF-kappaB activity. J Exp Med 201(1):105–115
60. Melvin A, Mudie S, Rocha S (2011) Further insights into the mechanism of hypoxia-induced NFκB. Cell Cycle 10(6):879–882
61. Flister MJ, Wilber A, Hall KL, Iwata C, Miyazono K, Nisato RE et al (2010) Inflammation induces lymphangiogenesis through up-regulation of VEGFR-3 mediated by NF-kappaB and Prox1. Blood 115(2):418–429
62. Illi B, Puri P, Morgante L, Capogrossi MC, Gaetano C (2000) Nuclear factor-kappaB and cAMP response element binding protein mediate opposite transcriptional effects on the Flk-1/KDR gene promoter. Circ Res 86(12):E110–E117

63. Xia C, Cheshire JK, Patel H, Woo P (1997) Cross-talk between transcription factors NF-kappa B and C/EBP in the transcriptional regulation of genes. Int J Biochem Cell Biol 29(12):1525–1539
64. Cappello C, Zwergal A, Kanclerski S, Haas SC, Kandemir JD, Huber R et al (2009) C/EBPbeta enhances NF-kappaB-associated signalling by reducing the level of IkappaB-alpha. Cell Signal 21(12):1918–1924
65. Hillier LW, Graves TA, Fulton RS, Fulton LA, Pepin KH, Minx P et al (2005) Generation and annotation of the DNA sequences of human chromosomes 2 and 4. Nature 434:724–731
66. Min Y, Ghose S, Boelte K, Li J, Yang L, Lin PC (2011) C/EBP-δ regulates VEGF-C autocrine signaling in lymphangiogenesis and metastasis of lung cancer through HIF-1α. Oncogene 30(49):4901–4909
67. Teng X, Li D, Catravas JD, Johns RA (2002) C/EBP-beta mediates iNOS induction by hypoxia in rat pulmonary microvascular smooth muscle cells. Circ Res 90(2):125–127
68. Liao S, Cheng G, Conner DA, Huang Y, Kucherlapati RS, Munn LL et al (2011) Impaired lymphatic contraction associated with immunosuppression. PNAS 108(46):18784–18789
69. Choi JS, Kim KB, Han W, Kim DS, Park JS, Lee JJ et al (2006) Efficacy of therapeutic angiogenesis by intramyocardial injection of pCK-VEGF165 in pigs. Ann Thorac Surg 82(2):679–686
70. Ferrarini M, Arsic N, Recchia FA, Zentilin L, Zacchigna S, Xu X et al (2006) Adeno-associated virus-mediated transduction of VEGF165 improves cardiac tissue viability and functional recovery after permanent coronary occlusion in conscious dogs. Circ Res 98(7):954–961
71. Bull DA, Bailey SH, Rentz JJ, Zebrack JS, Lee M, Litwin SE et al (2003) Effect of Terplex/VEGF-165 gene therapy on left ventricular function and structure following myocardial infarction. VEGF gene therapy for myocardial infarction. J Control Release 93(2):175–181
72. Pätilä T, Ikonen T, Rutanen J, Ahonen A, Lommi J, Lappalainen K et al (2006) Vascular endothelial growth factor C-induced collateral formation in a model of myocardial ischemia. J Heart Lung Transplant 25(2):206–213
73. Rutanen J, Rissanen TT, Markkanen JE, Gruchala M, Silvennoinen P, Kivelä A et al (2004) Adenoviral catheter-mediated intramyocardial gene transfer using the mature form of vascular endothelial growth factor-D induces transmural angiogenesis in porcine heart. Circulation 109(8):1029–1235
74. Saaristo A, Veikkola T, Enholm B, Hytönen M, Arola J, Pajusola K et al (2002) Adenoviral VEGF-C overexpression induces blood vessel enlargement, tortuosity, and leakiness but no sprouting angiogenesis in the skin or mucous membranes. FASEB J 16(9):1041–1049
75. Shay JE, Celeste SM (2012) Hypoxia-inducible factors: crosstalk between inflammation and metabolism. Semin Cell Dev Biol 23(4):389–394
76. Regenfu B, Bock F, Cursiefen C (2012) Corneal angiogenesis and lymphangiogenesis. Curr Opin Allergy Clin Immunol 12(5):548–554
77. Libby P (2012) Inflammation in atherosclerosis. Arterioscler Thromb Vasc Biol 32(9):2045–2051
78. Gordon JW, Shaw JA, Kirshenbaum LA (2011) Multiple facets of NF-κB in the heart: to be or not to NF-κB. Circ Res 108(9):1122–1132
79. Willerson JW, Ridker PM, Willerson JW, Ridker PM (2004) Inflammation as a cardiovascular risk factor. Circulation 109:II-2–II-1
80. Rickard AJ, Young MJ (2009) Corticosteroid receptors, macrophages and cardiovascular disease. J Mol Endocrinol 42(6):449–459
81. Libby P, Theroux P (2005) Pathophysiology of coronary artery disease. Circulation 111(25):3481–3488
82. Galea I, Beckmann I, Perry VH (2007) What is immune privilege (not)? Trends Immunol 28(1):12–18
83. Hori J, Vega JL, Masli S (2010) Review of ocular immune privilege in the year 2010: modifying the immune privilege of the eye. Ocul Immunol Inflamm 18(5):325–333

84. Yamagami S, Dana MR (2001) The critical role of lymph nodes in corneal alloimmunization and graft rejection. Invest Ophthalmol Vis Sci 42(6):1293–1298
85. Nykänen AI, Sandelin H, Krebs R, Keränen MA, Tuuminen R, Kärpänen T et al (2010) Targeting lymphatic vessel activation and CCL21 production by vascular endothelial growth factor receptor-3 inhibition has novel immunomodulatory and antiarteriosclerotic effects in cardiac allografts. Circulation 121(12):1413–1422
86. Syväranta S, Helske S, Lappalainen J, Kupari M, Kovanen PT (2012) Lymphangiogenesis in aortic valve stenosis—novel regulatory roles for valvular myofibroblasts and mast cells. Atherosclerosis 221(2):366–374
87. Steiner I, Krbal L, Dominik J (2010) Blood vessels and lymphatics in calcific aortic stenosis—in support of its inflammatory pathogenesis. Cesk Patol 46(2):33–36
88. Kholová I, Dragneva G, Cermáková P, Laidinen S, Kaskenpää N, Hazes T et al (2011) Lymphatic vasculature is increased in heart valves, ischaemic and inflamed hearts and in cholesterol-rich and calcified atherosclerotic lesions. Eur J Clin Invest 41(5):487–497
89. Xu X, Lu H, Lin H, Li X, Ni M, Sun H et al (2009) Aortic adventitial angiogenesis and lymphangiogenesis promote intimal inflammation and hyperplasia. Cardiovasc Pathol 18(5):269–278
90. Krishnan J, Suter M, Windak R, Krebs T, Felley A, Montessuit C et al (2009) Activation of a HIF1alpha-PPARgamma axis underlies the integration of glycolytic and lipid anabolic pathways in pathologic cardiac hypertrophy. Cell Metab 9(6):512–524
91. Nam EH, Park SR, Kim PH (2010) TGF-beta1 induces mouse dendritic cells to express VEGF and its receptor (Flt-1) under hypoxic conditions. Exp Mol Med 42(9):606–613
92. Mimura K, Kono K, Takahashi A, Kawaguchi Y, Fujii H (2006) Vascular endothelial growth factor inhibits the function of human mature dendritic cells mediated by VEGF receptor-2. Cancer Immunol Immunother 56(6):761–770
93. Hamrah P, Chen L, Zhang Q, Dana MR (2003) Novel expression of vascular endothelial growth factor receptor (VEGFR)-3 and VEGF-C on corneal dendritic cells. Am J Pathol 163(1):57–68
94. Mimura K, Kono K, Takahashi A, Kawaguchi Y, Fujii H (2007) Vascular endothelial growth factor inhibits the function of human mature dendritic cells mediated by VEGF receptor-2. Cancer Immunol Immunother 56(6):761–770
95. Nykänen AI, Sandelin H, Krebs R, Keränen MA, Tuuminen R, Kärpänen T et al (2010) Targeting lymphatic vessel activation and CCL21 production by vascular endothelial growth factor receptor-3 inhibition has novel immunomodulatory and antiarteriosclerotic effects in cardiac allografts. Circulation 121(12):1413–1422
96. Kotch LE, Iyer NV, Laughner E, Semenza GL (1999) Defective vascularization of HIF-1alpha-null embryos is not associated with VEGF deficiency but with mesenchymal cell death. Dev Biol 209(2):254–267
97. Wei H, Bedja D, Koitabashi N, Xing D, Chen J, Fox-Talbot K et al (2010) Endothelial expression of hypoxia-inducible factor 1 protects the murine heart and aorta from pressure overload by suppression of TGF-β signaling. PNAS 109(14):E841–E850
98. Negrini D, Moriondo A (2011) Lymphatic anatomy and biomechanics. J Physiol 589(12):2927–2934
99. Niessen K, Zhang G, Ridgway JB, Chen H, Kolumam G, Siebel CW et al (2011) The Notch1-Dll4 signaling pathway regulates mouse postnatal lymphatic development. Blood 118(7):1989–1997
100. Zheng W, Tammela T, Yamamoto M, Anisimov A, Holopainen T, Kaijalainen S et al (2011) Notch restricts lymphatic vessel sprouting induced by vascular endothelial growth factor. Blood 118(4):1154–1162
101. Cao R, Jensen LD, Söll I, Hauptmann G, Cao Y (2008) Hypoxia-induced retinal angiogenesis in zebrafish as a model to study retinopathy. PLoS One 3(7):e2748
102. Ota H, Katsube K, Ogawa J, Yanagishita M (2007) Hypoxia/Notch signaling in primary culture of rat lymphatic endothelial cells. FEBS Lett 581(27):5220–5226

103. Petrova TV, Karpanen T, Norrmén C, Mellor R, Tamakoshi T, Finegold D et al (2004) Defective valves and abnormal mural cell recruitment underlie lymphatic vascular failure in lymphedema distichiasis. Nat Med 10(9):974–981

104. Kume TJ (2012) The role of FoxC2 transcription factor in tumor angiogenesis. J Oncol 2012:204593, Epub 2011 Nov 16

Chapter 11
The Potential of the Epicardium to Act as a Source of Lymphatic Cells

Linda Klotz and Paul Riley

Abstract The lymphatic vasculature is a blind-ended network, crucial for tissue fluid homeostasis, immune surveillance, and lipid adsorption from the gut, as well as being the main route of cancer metastasis. The homeobox transcription factor *Prox1* is essential for the development of the lymphatic vasculature and is also expressed in the heart, where it is required for cardiomyocyte structure and growth as well as the cardiac conduction system. The cardiac lymphatic vasculature constitutes a largely overlooked but arguably very important part of the lymphatic system, which may have direct effects on cardiac morphogenesis, function, and response to injury. This chapter addresses whether the epicardium might act as a source of cells contributing to the cardiac lymphatics, by providing an overview of epicardial development and reflecting the current developmental potential of epicardium-derived progenitor cells. Furthermore, the morphogenesis of the coronary blood vasculature and controversy surrounding the origin of coronary vessels are elaborated to give a better understanding of endothelial development in the context of the heart. Finally, the specification of lymphatic endothelial cells and the development of the systemic lymphatic vasculature are described according to historical, venous versus non-venous origin, debate and linked to current knowledge on the potential for an epicardial contribution to the cardiac lymphatics.

Keywords Lymphatic • Heart • Prox1 • Epicardium • Development • Coronary vessel • Vasculogenesis • Angiogenesis • Epicardium-derived progenitor cell • Epithelial-to-mesenchymal transition • Lymphatic endothelial cell • Cardiovascular • Inflammation

L. Klotz
University College of London, Institute of Child Health, Molecular Medicine Unit, 30 Guilford Street, London WC1N 1EH UK

P. Riley (✉)
Department of Physiology, Anatomy, and Genetics, University of Oxford, Sherrington Building, South Parks Road, Oxford OX1 3PT, UK
e-mail: paul.riley@dpag.ox.ac.uk

G. Karunamuni (ed.), *The Cardiac Lymphatic System: An Overview*,
DOI 10.1007/978-1-4614-6774-8_11,

Introduction

Heart development begins approximately at embryonic day 7.0 (E7.0) in murine embryos and can be subdivided into three stages: early, mid-gestation, and late maturation stages. During this time, key events such as primary and secondary heart field development (cardiac progenitor cell specification), morphogenesis of the linear heart tube, cardiac looping and maturation, and growth and septation of the chambers take place. Beyond birth, further growth and maturation of the myocardium and other cardiac structures ensue [1]. The formation of the epicardium takes place during mid-gestational heart development, along with myocardial growth, endocardial cushion formation, ventricular septation and trabeculation, as well as coronary and lymphatic vessel development [2–4]. This chapter will provide a brief overview of the development of the epicardium, the coronary blood vasculature, and the specification and development of the systemic and cardiac lymphatic vasculature, and also discuss the possibility of whether the epicardium may play a role in cardiac lymphatic development.

The Epicardium

The chambers of the mature vertebrate heart consist of three tissue layers: the myocardium and the endocardium and epicardium, which line the myocardium on the inside and outside, respectively. The epicardium comprises a single-cell epithelial sheet and is connected to the myocardium by subepicardial connective tissue. The epicardium undergoes epithelial-to-mesenchymal transition (EMT) during heart development and during postnatal repair and regeneration [5] to give rise to a population of so-called epicardium-derived progenitor cells (EPDCs). These EPDCs invade the myocardium and differentiate into various cell types, including cardiac fibroblasts, smooth muscle cells, pericytes, and cardiomyocytes, and may also play a role in Purkinje fiber development [6, 7]. There is an accumulation of evidence to indicate that the interactions between the epicardium and myocardium are crucial for correct differentiation and patterning of the heart. It has been shown that secondary cardiac defects, for example, thinning of the myocardium or interventricular septal defects, are consequences of genetic or surgical perturbation of epicardial development [8–18].

Development of the Epicardium

The epicardium and its derivatives were originally thought to develop from the myocardium until the late 1960s, when this theory was disproven [19]. The true origin of the epicardium was not discovered until decades later, when research in multiple vertebrate model organisms demonstrated an extracardiac source of cells

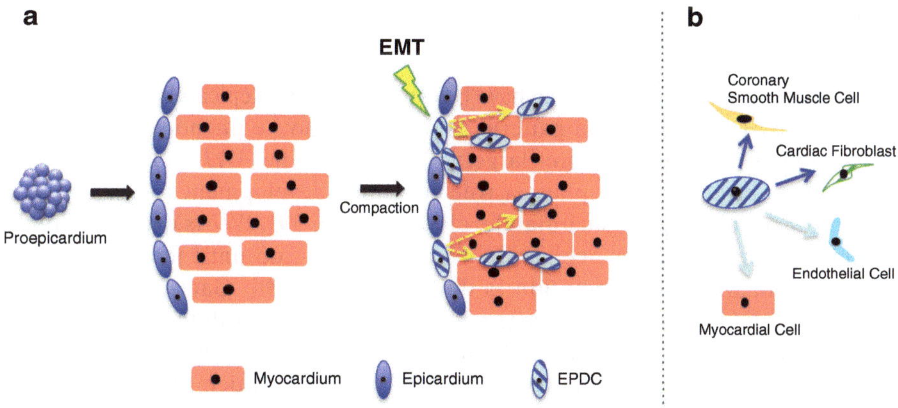

Fig. 11.1 (**a**) Proepicardial cells migrate towards the myocardium, where they attach to form a thin epithelial monolayer, the epicardium. A subset of epicardial cells undergoes epithelial-to-mesenchymal transition (EMT) to form epicardium-derived progenitor cells (EPDCs). EPDCs are induced to migrate into the subepicardial space and myocardium, leading to compaction of the myocardium, and subsequently differentiate into different cell types. (**b**) EPDCs can differentiate into coronary smooth muscle cells and interstitial cardiac and adventitial fibroblasts, with ongoing debate concerning their ability to differentiate into coronary endothelial or myocardial cells

giving rise to the epicardium: the proepicardium (PE), a cauliflower-shaped mass of coelomic cells which protrudes towards the pericardial cavity [20–22].

The proepicardium can first be visualized in mice around E8.5 [23] and is made up of a mesodermal cell population which protrudes immediately posterior to the sinoatrium in close proximity to the liver [20, 24–28]. The population is composed of at least two cell types: an external mesothelial epithelium and a mesenchymal core rich in extracellular matrix [28]. The proepicardium develops bilaterally symmetric in mice and zebra fish with two cell clusters that merge together to form a single proepicardium [22, 23, 27], whereas the chick proepicardium develops asymmetrically, with the right anlage appearing at HH-stage 14 [29] and the left anlage appearing at HH-stage 15/16. In chick embryos, only the right proepicardial anlage matures [30, 31]. Formation of the epicardial sheet requires the attachment of proepicardial cells to the naked myocardium. Proepicardial cells migrate towards the myocardial surface and subsequently flatten and spread to form a thin epithelial monolayer and a matrix-rich subepicardial space [20, 25, 26] (Fig. 11.1a). Starting in the atrioventricular groove, the epicardium grows to completely engulf the heart in a fixed spatiotemporal pattern as development progresses. The fate of the PE has been extensively studied in avian embryos using chick-quail chimeras and retroviral labeling techniques. The PE gives rise to most of the epicardium, with the exception of a small population near the outflow tract, which arises from the coelomic pericardial epithelium near the aortic sac [32]. Besides the primitive epicardium, the proepicardium also gives rise to blood endothelial cells; smooth muscle cells of the coronary blood vessels; perivascular, intermyocardial, and subendocardial fibroblasts; and a small number of endocardial cells [10, 33–37].

Epicardial Genes

There are a number of genes which have been associated with epicardial development, mostly based on expression patterns and mutant mouse phenotypes; however, their downstream targets are largely unknown. The *Wt1* gene encodes a zinc-finger protein which plays a role in tumor development [38] as well as kidney, urogenital, and epicardial development [15, 39]. *Wt1* is expressed in the primitive proepicardium-, epicardium-, and EPDCs, up until the point of differentiation, when its expression is downregulated [40]. Gata4, another zinc-finger protein, is expressed in the proepicardium and epicardium [41, 42], along with chicken Gata5 (cGATA5), which has been used to drive Cre-recombinase expression ectopically in the epicardium of transgenic mice [43]. GATA transcription factors also play a critical role during coronary artery formation [44]. Two T-box genes, *Tbx5* and *Tbx18*, are expressed during epicardial development [45]. The expression of *Tbx5* is transient and becomes lost after the epicardium has fully formed, while *Tbx18* is strongly expressed in the proepicardium and epicardium [30], although recent evidence suggests that *Tbx18* is dispensable for epicardial development [46]. Members of the forkhead transcription factor family, *FoxC1* and *FoxC2*, are expressed in a subset of cells in the proepicardium [47] and are also involved in kidney, cardiovascular, corneal, and cerebellar development and somitogenesis [48–51]. Interestingly, these transcription factors also play a role in lymphatic development [51, 52]. Another protein involved in both epicardial and lymphatic development is the transmembrane glycoprotein podoplanin [53, 54].

Deletion of key genes necessary for proepicardial and epicardial development have shown severe defects in cardiomyoblast proliferation as well as coronary vascular development [13–15, 42, 53, 55], although the effect on cardiac lymphatics has been entirely overlooked in this respect. Thus, the epicardium contributes to the development of key cardiac structures either by acting as a source of cell types or by providing signals, which will be described in more detail concerning coronary vasculogenesis, to direct the differentiation of other cells into distinct lineages. However, it is important to note that fate-mapping studies using epicardial marker genes for lineage trace analysis have proven to be difficult, as many of those genes, or the specific transgenic reporter/Cre constructs targeted to these loci, are also expressed in other tissues (earlier or later) during development (e.g., *Wt1* [56], *Tbx18* [57], *Gata5* [58]). This has made accurate lineage tracing, via Cre-lox strategies, of epicardial cell fate during development problematic. Furthermore, the timing and accuracy of Cre-mediated recombination depends not only on the expression level of the gene driving Cre activity but also on the sensitivity of the reporter construct itself [59]. Therefore, confirming the epicardial origin of a cell or tissue type requires extremely careful analysis, utilizing multiple approaches in parallel with lineage tracing to include accurate marker expression and conventional embryology (dye labeling, organ culture, etc.).

Epicardium-Derived Progenitor Cells

Once the PE cells have formed the primitive epicardium, a subset of epicardial epithelial cells undergo EMT to transform into a population of highly migratory, invasive cells called EPDCs [34], which then migrate into the subepicardial space [20] (Fig. 11.1a). It is currently still unclear whether all cells within the epicardium have the potential to undergo EMT and form EPDCs, selected on the basis of environmental cues or cell–cell interactions, or whether epicardial cells are heterogeneous and, therefore, only a subset are competent to undergo EMT [6, 60]. Recent evidence points towards a heterogeneous population in origin as well as function [61]. EMT happens throughout development to provide mesenchyme to developing embryonic structures, and in the heart it is observed at the atrioventricular junction, in the ventricular (but not atrial) epicardium, and at the junction between the ventricles and the outflow tract [62]. In mice, once EPDCs migrate into the myocardium, a thick, compact myocardial layer develops, which is more pronounced in the left ventricle [6]. The close reciprocal interactions between the myocardium and epicardium mean that any disturbance of epicardial outgrowth and EMT and migration of EPDCs leads to severe hypoplasia of the compact myocardium. This was shown via both mechanical inhibition of epicardial outgrowth [16] as well as in mouse mutants lacking key genes involved in these processes, including *PDGFRα*(alpha), *podoplanin*, and *Wt1* [53, 63–65]. It is generally accepted that EPDCs can differentiate into interstitial cardiac fibroblasts, adventitial fibroblasts, and coronary smooth muscle cells, confirmed by a series of seminal avian retroviral labeling and chick-quail chimera studies [33, 34, 37, 66, 67] (Fig. 11.1b) as well as, more recently, transgenic mouse lineage tracing studies [27, 33, 34, 37, 58, 68, 69]. There is some controversy and ongoing debate about the potential of EPDCs to differentiate into myocardial cells [70, 71] and coronary endothelium [6], of which the latter will be discussed in more detail in the next section.

Development of Coronary Blood Vessels

In the past three decades, there has been substantial progress in the understanding as to how systemic blood vessels develop as well as functional insight into the factors controlling their development [72]. Blood vessel development requires the tight coordination of cell proliferation, differentiation, migration, matrix adhesion, and cell–cell signaling [73]. It can be divided into vasculogenesis and angiogenesis. Vasculogenesis refers to the de novo formation of blood vessels via differentiation of angioblast or hemangioblast precursors, while angiogenesis denominates vascular sprouting or remodeling from existing vessels, commonly occurring throughout vascular development. It is thought that the primary capillary plexus is formed by vasculogenesis (e.g., the dorsal aorta and cardinal vein [74]) and later remodeled by angiogenesis [75, 76].

Although coronary blood vessels have been studied for over a century, comparatively little is known about the origin of cells and the underlying developmental programs of arguably one of the most important vascular networks. Simplistically, the heart pumps blood around the body through arteries, into smaller arterioles and even smaller capillary beds, which form vast plexi to exchange gases and metabolic products—only to be returned through venules and veins and then pumped to the lungs, where the blood is replenished with oxygen. The heart muscle continuously contracts to empty (diastole) and fill (systole) during the lifetime of adult mammals, and for this to work effectively, the heart is in constant need of its own supply of oxygen and nutrients from a fully functioning coronary vessel network. This requirement is consistent with mid-gestational development, when organ growth means that diffusion alone can no longer adequately support heart function.

Signaling Between the Epicardium and Coronary Vessels

Development of the coronary vasculature begins with the formation of a vascular network which is subsequently remodeled to give rise to the mature coronary tree [77–79]. This takes place in chick embryos from HH23 onwards [78] and in mouse embryos from E11.5 onwards [79, 80] and proceeds in four stages: migration of endothelial precursors, formation of an endothelial plexus, remodeling of the endothelial plexus, and growth and maturation of the coronary vessels [77]. While coronary vasculogenesis is largely complete by mid-gestation, the remodeling via angiogenesis and arteriogenesis continues throughout development and postnatally [81].

The development of the coronary vessels is closely associated with the development of the epicardium, which acts as a signaling center via multiple canonical pathways (e.g., retinoic acid (RA), fibroblast growth factor (FGF), hedgehog (HH), Wnt, and Notch) [60]. As mentioned previously, deletion of a number of genes necessary for epicardial development in the mouse has resulted in severe defects in cardiomyoblast proliferation and coronary vascular development [13–15, 42, 53, 55]. In *Fog2* (Friend-of Gata2) mutants, while the epicardium itself appears normal, there is a failure to undergo epicardial EMT, leading to severe defects in myocardial proliferation and coronary vessel development [44, 82]. Epicardial-derived FGF signals are also important for coronary vascular development, in addition to mediating myocardial growth [83], and have been shown to indirectly regulate coronary vascular plexus formation via HH signaling. Conditional knockouts of FGF9 and FGF-R1/FGF-R2 in mice results in delayed epicardial hedgehog expression and delayed coronary plexus formation due to the negative effect on downstream *Vegf-A*, *Vegf-B*, *Vegf-C*, and *Ang2* expression [80]. Two recent studies have suggested that the Notch pathway is also involved in the development of mouse epicardium and coronary vessels, as conditional mutants of Notch1 and its downstream effector, RBPJk, displayed severely reduced or abnormally dilated coronary vessels, respectively [84, 85].

Origin of Coronary Blood Endothelial Cells

For centuries the origin of coronary blood vessels was uncertain, and the assumption that they arise from endothelial sprouts originating in the aortic sinuses, myocardial sinusoids, and the sinus venosus was first made in the early twentieth century [86–88]. In the late 1980s, attempts to refute this assumption using descriptive data suggested that coronary arteries develop by ingrowth of a subepicardial endothelial plexus into the aorta [89–91]. Later quail-chick chimera studies on the one hand confirmed this hypothesis [67] and on the other hand demonstrated that cells of the coronary blood vessels come from a primarily extracardiac progenitor cell population, namely, the proepicardium [33–35, 92]. However, it is not clear whether quail angioblasts transplanted with the proepicardium differentiated there or whether they represented a migratory cell population (from a tissue such as the liver) that passed through the proepicardium en route to the subepicardium. Thus, there are two schools of thought dividing the field, with more recent mouse studies supporting both views.

Merki and colleagues reported no contribution of the proepicardial lineage to the coronary endothelium using cGata5-Cre transgenic mice [58]. A study by Red-Horse and colleagues demonstrated that coronary vessels arise from venous endothelial cells sprouting off the sinus venosus around E11.5, which dedifferentiate and are reprogrammed into arteries, capillaries, and veins [79]. However, this study did not exclude the possibility that endothelial cells may contribute to the coronary vasculature by coming from a distinct source (proepicardium, endocardium) or by arising as earlier progenitors elsewhere and then migrating into, or through, the sinus venosus [93]. This critique is further supported by findings from Katz and colleagues, who identified proepicardial markers *Scleraxis* and *Semaphorin 3D* which delineate previously uncharacterized proepicardial subcompartments. They demonstrated that the proepicardium itself is heterogeneous and has distinct subpopulations of cells, some of which can give rise to coronary endothelial cells via the sinus venosus [61].

Thus, the debate about the origin of coronary endothelial cells is still ongoing, with much remaining to be understood about the proepicardium as a cellular source of coronary endothelial cells.

As the development of the lymphatic vasculature occurs after the cardiovascular system and the first lymphatic vessels sprout from the cardinal vein, it is pertinent to acknowledge the close relationship between these two vascular networks. Understanding the connection between blood vessel and lymphatic vessel growth may lead to a greater insight into lymphatic development, particularly in organs where it has not been studied extensively.

Lymphatic Vessel Development

The lymphatic vasculature is a blind-ended network covering most tissues and organs of the body and is an essential component of vertebrate development and homeostasis. The lymphatic vasculature is crucial for tissue fluid regulation, immune surveillance, and lipid adsorption from the gut and also serves as the main route for cancer metastasis. A number of human diseases are associated with reduced lymphatic vascular function or aberrant growth and development of lymphatic vessels. Problems in the lymphatic system can lead to pathologies such as lymphedema, delayed immune response, obesity, and hypertension [94].

The Historical Debate: Venous Versus Mesenchymal Origin?

The first published studies of the lymphatic vasculature date back to the mid-seventeenth century, when Gaspare Aselli, an Italian professor of anatomy and surgery, described the vessels he observed in a dog's abdomen as "milky veins." [95] Although further descriptive studies ensued, the origin of the lymphatic vessels remained unclear. In 1902, Florence Sabin proposed what remains to this day the most widely accepted model of lymphatic development. Following ink injection experiments performed on pig embryos, Sabin postulated that primary lymph sacs originate from endothelial cells which bud off from the veins [96]. This model suggests that the entire peripheral lymphatic vasculature develops by endothelial sprouting and remodeling from the primary lymph sacs, also known as the centrifugal theory. A few years later, an alternative, less popular model of lymphatic development was proposed by Huntington and McClure, who suggested that lymph sacs arise in the mesenchyme via distinct progenitor cells independently of veins, and later secondarily establish venous connections—the centripetal theory [97]. Although both models were proposed over a century ago, the discussion is still a current topic of interest, and more recent evidence has supported both the venous [98–101] as well as a distinct mesenchyme-derived lymphatic vasculature in various vertebrate model organisms [102–106].

Specification of Lymphatic Endothelial Cells

The specification of murine lymphatic endothelial cells (LECs) takes places during mid-gestation. The lymphatic vasculature begins to develop after the blood vasculature, and the cells that commit to a lymphatic identity originate within the dorsolateral walls of the cardinal veins. In the mouse at around embryonic day 9.5 (E9.5), a subset of COUP-TFII$^+$ venous endothelial cells in the cardinal vein begin to express transcription factors *Sox18* and *Prox1*, and at that stage are classified as LEC progenitor cells [96, 98, 107–110]. *Sox18* binds directly to the proximal promoter of *Prox1* to activate its expression; however, *Sox18* expression is transient and as such only detected in LECs and newly forming vessels up to ~E14.5 [108]. Nevertheless, its role in developmental lymphangiogenesis is crucial as *Sox18*$^{-/-}$ embryos show a complete blockade of LEC differentiation from the cardinal vein [108]. This is consistent with *Prox1*-deficient embryos, which are devoid of lymphatic vessels due to a failure of LEC specification. Although sprouting from the veins is still observed, these *Prox1*-mutant cells have blood rather than lymphatic endothelial characteristics [109, 111].

Once specified in wild-type embryos, LECs subsequently delaminate from the walls of the anterior cardinal vein and migrate into the mesenchyme in a polarized manner between E10 and E11.5 to form primitive lymph sacs [111] (Fig. 11.2). Lymph sacs are the embryonic structures from which most of the lymphatic vasculature is eventually derived. The sprouting, proliferation, and migration of LECs away from the cardinal vein are regulated by Vegf-C through its receptor VEGFR-3, and it has been shown that sprouting of Prox1^{+} LECs in the cardinal vein is inhibited in Vegf-C$^{-/-}$ mutants [112, 113]. The lymphatic vasculature and blood vasculature are completely separated, with the exception of a few key connections where lymph is emptied back into the venous system at the junction of the thoracic duct and subclavian vein; in the renal, hepatic, and adrenal veins; and in the lymph nodes [114]. The blood-lymphatic separation is regulated by platelet aggregation at the junction sites between the cardinal vein and lymph sacs, induced by stimulation of the CLEC-2 receptor in platelets via podoplanin from LECs [115, 116]. The development and remodeling of the systemic lymphatic vasculature is complete by 2 weeks after birth [117]. Recent evidence has suggested that lymphatic vessels not only sprout from the cardinal vein but also from intersomitic veins [100] (Fig. 11.2), where the same pathways underlie lymphatic specification and migration. This broadens the scope of the developmental origin of lymphatic vessels, albeit still in support of the venous origin model.

Origins of Lymphatic Endothelial Cells

There are a number of different cell types, other than venous endothelial cells, which have been implicated in physically contributing to newly forming lymphatic vessels—interestingly, this only applies to lymphangiogenesis under pathological conditions or postnatally, rather than during embryonic development.

Pronounced neo-lymphangiogenesis has been found to be involved with the pathogenesis of organ rejection, psoriasis, airway inflammation, and rheumatoid arthritis [118–122], and NF-κB and LTα have recently been implicated in driving *Prox1* expression to initiate lymphangiogenesis during inflammation [123, 124]. However, the origin of cells contributing to newly formed vessels in pathological settings is ambiguous. Circulating precursors of LECs in mammals have been shown to differentiate from hematopoietic tissue-derived endothelial progenitors [125] which bear Syk- and Slp-76 antigens [126], from transdifferentiating leukocytes [106], or from transdifferentiating Cd11b^{+} macrophages [118, 127]. In addition, other studies have reported bone marrow-derived cells as a potential source, given their ability to acquire LEC antigens during lymphangiogenesis, both during normal homeostasis and under pathological conditions [128–130].

Recent findings have suggested that increasing lymphangiogenesis and stimulating lymphatic vessel function leads to a reduction in chronic skin inflammation [131, 132]; therefore understanding the origin of cells contributing to new vessels is crucial. In contrast, the origin of progenitor cells contributing to the cardiac lymphatic vasculature is entirely unknown. Furthermore, the role of the lymphatic

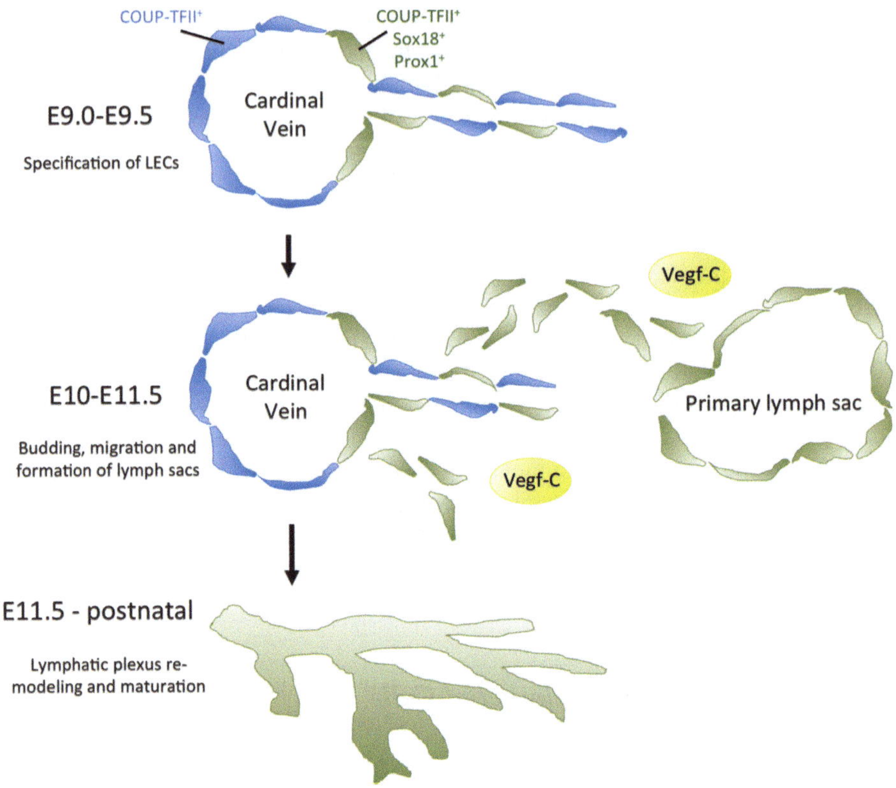

Fig. 11.2 The development of the murine lymphatic system. In mice, the LEC competence factor Sox18 starts to be expressed in a subpopulation of COUP-TFII$^+$ venous endothelial cells around E9.5. The commitment towards LEC differentiation is initiated upon Prox1 expression. Vegf-C signaling via Vegfr-3 induces Prox1-expressing cells to migrate away from the cardinal vein and intersomitic veins, by budding off to form the first primitive lymph sacs. Later, the whole lymphatic vascular network develops from these primitive lymph sacs and undergoes remodeling to form mature capillaries and vessels

system during cardiac inflammation is unknown, and in the context of adult cardiac injury (e.g., myocardial infarction), lymphatic vessel clearance of the inflammatory infiltrate may significantly improve cardiac outcome. A recent study demonstrated an increased remodeling of lymphatic vessels surrounding the infarct area; however, neither the cellular contribution and mechanisms underlying this response nor the potential functional benefits were examined [133].

The Cardiac Lymphatic Vasculature

The major focus thus far has been on the systemic lymphatic vasculature, while organ-based lymphatics have been largely overlooked. In particular, the lymphatic system in the heart has not been studied in great detail, and the mechanisms regulating cardiac lymphatic development are yet to be described. Although recent publications describe the expression of lymphatic markers in developing murine and chick hearts [3, 4, 134], it is unclear where these LECs come from and whether other tissues may play a role in cardiac lymphatic development. The knowledge of the contribution of cardiac tissues will be beneficial to understanding the interplay between the heart and cardiac lymphatic vasculature, both during development and in disease.

Prox1 in the Heart

Prospero was originally isolated from *Drosophila* in 1991, where it was found to be fundamental for specifying neuronal fate [135, 136]. It was not until 1993 that *Prox1* (*prospero-related homeobox gene1*) was cloned in mice [137], and later its role in mammalian hematopoietic stem cell, retina, lens, and lymphatic development emerged [111, 138, 139]. The specific role in lymphatic development has been corroborated in other model organisms, such as chick [140], *Xenopus* [141], and zebra fish [141], revealing *Prox1* function to be highly conserved throughout evolution. Further studies have revealed that *Prox1* is the "master regulator" of lymphatic development [142], whose expression must be maintained within cells to retain a lymphatic identity and prevent them from being reprogrammed into blood endothelial cells (BECs) [143]. In this way, *Prox1* acts as a binary molecular switch, turning the BEC program "off" and the LEC program "on" [144].

In addition to being essential for the development of the lymphatic system, *Prox1* is also expressed in the heart [137] and has been implicated in maintaining cardiomyocyte structure and growth during heart development [145], and more recently in maintaining the adult cardiac conduction system [146]. Knocking out *Prox1* conditionally in cardiomyocytes has shown that *Prox1* is required for direct transcriptional regulation of the genes encoding structural proteins α-actinin, N-RAP, and zyxin, which collectively function to maintain association of the sarcomere [145]. Cardiomyocyte-specific *Prox1* knockout mice not only show abnormal heart development and defective sarcomere integrity compared to wild-type mice but also varying severities of edema, which is indicative of problems in the lymphatic system. Despite a wider understanding of the function of *Prox1* within the developing heart, the associated signaling pathways are still unknown.

The Role of the Epicardium During Cardiac Lymphatic Development

What is known thus far about the possible contribution of the proepicardium to coronary lymphatic vessels has largely arisen from chick-quail transplantation experiments. A study by Wilting and colleagues, in which a quail proepicardium was transplanted into a chick host, demonstrated that avian cardiac lymphatics do not develop from the proepicardium [134]. Although they show donor-derived blood vessels in the chick host, the lymphatic vasculature is completely derived from the host itself, with the exception of one large lymphatic trunk at the base of the heart, attributed as a lympho-venous anastomosis, and thus possible homing of donor-derived venous endothelial cells into the Prox1$^+$ lymphatic vessel. These results not only suggest that cardiac lymphatic vessels do not arise from the proepicardium or epicardium, but further imply that the cardiac lymphatic vessels do not arise from "local" venous endothelial cells (coronary veins) either, which themselves were donor-derived. As mentioned earlier, the current widely accepted dogma in lymphatic development is that all lymphatic vessels are derived from the cardinal vein and intersomitic veins [98, 100], which then sprout to give rise to all peripheral lymphatics. The findings from the chick-quail transplantation experiments would appear to support this theory.

It is significant to note that the aforementioned avian study is the first and only one of its kind investigating the contribution of the proepicardium to cardiac lymphatic vessels. As has been shown in the field of coronary vessel development, the discussion about the origin of endothelial cells in the heart is ongoing, specifically in reference to the variation seen between different model organisms and confounding species differences. Therefore, more investigation into this area of research needs to be performed to gain a deeper insight, especially given the relative paucity of appropriate mammalian models. A recent study showing different anatomical origins of vessels in the murine heart also emphasizes the possible heterogeneity of cells underlying cardiac lymphatic vessels [147]. There are a number of potential sources of progenitor cells that could contribute to the cardiac lymphatics. Although dermal lymphatics have been shown not to be derived from macrophages during development [148], macrophages are highly abundant during heart development [149] and therefore a candidate source for cardiac lymphatic development, which should be further investigated.

Despite transcription factors COUP-TFII, Sox18, and Prox1 being indispensable for lymphatic vascular development [150], there are a number of other pathways equally important for the correct development and remodeling of lymphatic vessels. Multiple studies have shown evidence for FGF signaling as being crucial for lymphatic development during embryogenesis and tumor vascularization and in ex vivo primary cell cultures. Shin and colleagues identified FGFR-3 as a direct target of Prox1 [151] and found it to be upregulated in LECs and lymphatic vessels throughout development. Further studies have shown that FGF-2-/FGFR-1-mediated lymphangiogenesis is dependent on Vegfr-3/Vegf-C signaling [152–155]. Retinoic acid has also been implicated in this pathway upstream of FGF receptor signaling [156]. In addition, classical Notch signaling has been shown to repress *Prox1* and *COUP-TFII* expression and the induction of LEC fate in vitro [157]. Interestingly, zebra fish and mouse models provide conflicting evidence suggesting that Notch signaling either plays a positive regulatory role or is not required at all for lymphangiogenesis [110, 158].

Conclusion

The development of the epicardium is a key process playing a central role during heart development. One factor complicating the study of epicardial progenitor cell origin and potential, and further confounded by current mouse models, is the lack of epicardial or EPDC-specific genes [159]. It is clear that the formation of the epicardium, as well as signaling between the epicardium and myocardium, is important for the correct development of the vessels and muscle of the embryonic heart, which in turn may also affect the morphogenesis and function of the cardiac lymphatic network. The aforementioned pathways involved in lymphatic development are also associated with epicardial development in conjunction with the coronary vasculature, and circumstantial evidence exists for a connection between lymphatic development and epicardial cell contribution to the heart. Therefore, although studies in the chick suggest that the epicardium does not directly contribute cells to the cardiac lymphatic vasculature, the emergence of new tools to study the epicardial lineage may assist in determining whether this is also true in mammals, including man. In addition, the adult epicardium can be induced to proliferate, migrate, and differentiate into cardiovascular derivatives following injury or in disease models [70, 160, 161], via the upregulation of an embryonic gene program. The possibility of an epicardial contribution to cardiac lymphatic vessels in a cardiovascular disease setting remains a tangible prospect, and one that is subject to ongoing investigation.

Acknowledgements LK is funded by a Wellcome Trust 4-year PhD studentship (Ref: 089592/Z/09/A). PRR is funded by a British Heart Foundation Personal Chair Award (Ref: CH/11/1/28798).

References

1. Rosenthal N, Harvey RP (2010) Heart development and regeneration, 1st edn. Academic, San Diego
2. Gittenberger-De Groot AC, Bartelings MM, Deruiter MC, Poelmann RE (2005) Basics of cardiac development for the understanding of congenital heart malformations. Pediatr Res 57(2):169–176
3. Juszyn ski M, Ciszek B, Stachurska E, Jabłon ska A, Ratajska A (2008) Development of lymphatic vessels in mouse embryonic and early postnatal hearts. Dev Dyn 237(10):2973–2986
4. Karunamuni G, Yang K, Doughman Y-Q, Wikenheiser JC, Bader D, Barnett J, Austin A, Parsons-Wingerter P, Watanabe M (2010) Expression of lymphatic markers during avian and mouse cardiogenesis. Anat Rec (Hoboken) 293(2):259–270
5. Zhou B, Pu WT (2011) Epicardial epithelial-to-mesenchymal transition in injured heart. J Cell Mol Med 15(12):2781–2783
6. Gittenberger-De Groot AC, Winter EM, Bartelings MM, Jose Goumans M, Deruiter MC, Poelmann RE (2012) The arterial and cardiac epicardium in development, disease and repair. Differentiation 84:41–53
7. Smart N, Dubé KN, Riley PR (2009) Coronary vessel development and insight towards neovascular therapy. Int J Exp Pathol 90(3):262–283

8. Wilson JG, Warkany J (1949) Aortic-arch and cardiac anomalies in the offspring of vitamin A deficient rats. Am J Anat 85(1):113–155
9. Kreidberg JA, Sariola H, Loring JM, Maeda M, Pelletier J, Housman D, Jaenisch R (1993) WT-1 is required for early kidney development. Cell 74(4):679–691
10. Männer J (1993) Experimental study on the formation of the epicardium in chick embryos. Anat Embryol 187(3):281–289
11. Kastner P, Grondona JM, Mark M, Gansmuller A, LeMeur M, Decimo D, Vonesch JL, Dollé P, Chambon P (1994) Genetic analysis of RXR alpha developmental function: convergence of RXR and RAR signaling pathways in heart and eye morphogenesis. Cell 78(6):987–1003
12. Sucov HM, Dyson E, Gumeringer CL, Price J, Chien KR, Evans RM (1994) RXR alpha mutant mice establish a genetic basis for vitamin A signaling in heart morphogenesis. Genes Dev 8(9):1007–1018
13. Kwee L, Baldwin HS, Shen HM, Stewart CL, Buck C, Buck CA, Labow MA (1995) Defective development of the embryonic and extraembryonic circulatory systems in vascular cell adhesion molecule (VCAM-1) deficient mice. Development 121(2):489–503
14. Yang JT, Rayburn H, Hynes RO (1995) Cell adhesion events mediated by alpha 4 integrins are essential in placental and cardiac development. Development 121(2):549–560
15. Moore AW, McInnes L, Kreidberg J, Hastie ND, Schedl A (1999) YAC complementation shows a requirement for Wt1 in the development of epicardium, adrenal gland and throughout nephrogenesis. Development 126(9):1845–1857
16. Gittenberger-de Groot AC, Vrancken Peeters MP, Bergwerff M, Mentink MM, Poelmann RE (2000) Epicardial outgrowth inhibition leads to compensatory mesothelial outflow tract collar and abnormal cardiac septation and coronary formation. Circ Res 87(11):969–971
17. Pennisi DJ, Ballard VLT, Mikawa T (2003) Epicardium is required for the full rate of myocyte proliferation and levels of expression of myocyte mitogenic factors FGF2 and its receptor, FGFR-1, but not for transmural myocardial patterning in the embryonic chick heart. Dev Dyn 228(2):161–172
18. Männer J, Schlueter J, Brand T (2005) Experimental analyses of the function of the proepicardium using a new microsurgical procedure to induce loss-of-proepicardial-function in chick embryos. Dev Dyn 233(4):1454–1463
19. Manasek FJ (1969) Embryonic development of the heart. II. Formation of the epicardium. J Embryol Exp Morphol 22(3):333–348
20. Viragh S, Challice CE (1981) The origin of the epicardium and the embryonic myocardial circulation in the mouse. Anat Rec (Hoboken) 201(1):157–168
21. Männer J (1992) The development of pericardial villi in the chick embryo. Anat Embryol 186(4):379–385
22. Serluca FC (2008) Development of the proepicardial organ in the zebrafish. Dev Biol 315(1):18–27
23. Schulte I, Schlueter J, Abu-Issa R, Brand T, Männer J (2007) Morphological and molecular left–right asymmetries in the development of the proepicardium: a comparative analysis on mouse and chick embryos. Dev Dyn 236(3):684–695
24. Ho E, Shimada Y (1978) Formation of the epicardium studied with the scanning electron microscope. Dev Biol 66(2):579–585
25. Hiruma T, Hirakow R (1989) Epicardial formation in embryonic chick heart: computer-aided reconstruction, scanning, and transmission electron microscope studies. Am J Anat 184(2):129–138
26. Viragh S, Gittenberger-de Groot AC, Poelmann RE, Kalman F (1993) Early development of quail heart epicardium and associated vascular and glandular structures. Anat Embryol 188(4):381–393
27. Männer J, Pérez-Pomares JM, Macias D, Munoz-Chápuli R (2001) The origin, formation and developmental significance of the epicardium: a review. Cells Tissues Organs 169(2):89–103

28. Nahirney PC, Mikawa T, Fischman DA (2003) Evidence for an extracellular matrix bridge guiding proepicardial cell migration to the myocardium of chick embryos. Dev Dyn 227(4):511–523
29. Hamburger V, Hamilton HL (1951) A series of normal stages in the development of the chick embryo. J Morphol 88(1):49–92
30. Schlueter J, Männer J, Brand T (2006) BMP is an important regulator of proepicardial identity in the chick embryo. Dev Biol 295(2):546–558
31. Schlueter J, Brand T (2009) A right-sided pathway involving FGF8/Snail controls asymmetric development of the proepicardium in the chick embryo. Proc Natl Acad Sci USA 106(18):7485–7490
32. Perez-Pomares JM, Phelps A, Sedmerova M, Wessels A (2003) Epicardial-like cells on the distal arterial end of the cardiac outflow tract do not derive from the proepicardium but are derivatives of the cephalic pericardium. Dev Dyn 227(1):56–68
33. Dettman RW, Denetclaw W, Ordahl CP, Bristow J (1998) Common epicardial origin of coronary vascular smooth muscle, perivascular fibroblasts, and intermyocardial fibroblasts in the avian heart. Dev Biol 193(2):169–181
34. Gittenberger-de Groot AC, Vrancken Peeters MP, Mentink MM, Gourdie RG, Poelmann RE (1998) Epicardium-derived cells contribute a novel population to the myocardial wall and the atrioventricular cushions. Circ Res 82(10):1043–1052
35. Pérez-Pomares J-M, Carmona R, González-Iriarte M, Atencia G, Wessels A, Muñoz-Chápuli R (2002) Origin of coronary endothelial cells from epicardial mesothelium in avian embryos. Int J Dev Biol 46(8):1005–1013
36. Mikawa T, Fischman DA (1992) Retroviral analysis of cardiac morphogenesis: discontinuous formation of coronary vessels. Proc Natl Acad Sci USA 89(20):9504
37. Mikawa T, Gourdie RG (1996) Pericardial mesoderm generates a population of coronary smooth muscle cells migrating into the heart along with ingrowth of the epicardial organ. Dev Biol 174(2):221–232
38. Haber DA, Buckler AJ, Glaser T, Call KM, Pelletier J, Sohn RL, Douglass EC, Housman DE (1990) An internal deletion within an 11p13 zinc finger gene contributes to the development of Wilms' tumor. Cell 61(7):1257–1269
39. Buckler AJ, Pelletier J, Haber DA, Glaser T, Housman DE (1991) Isolation, characterization, and expression of the murine Wilms' tumor gene (WT1) during kidney development. Mol Cell Biol 11(3):1707
40. Pérez-Pomares JM, Phelps A, Sedmerova M, Carmona R, González-Iriarte M, Munoz-Chápuli R, Wessels A (2002) Experimental studies on the spatiotemporal expression of WT1 and RALDH2 in the embryonic avian heart: a model for the regulation of myocardial and valvuloseptal development by epicardially derived cells (EPDCs). Dev Biol 247(2):307–326
41. Nemer G, Nemer M (2003) Transcriptional activation of BMP-4 and regulation of mammalian organogenesis by GATA-4 and -6. Dev Biol 254(1):131–148
42. Watt A, Battle MA, Li J, Duncan SA (2004) GATA4 is essential for formation of the proepicardium and regulates cardiogenesis. Proc Natl Acad Sci 101(34):12573–12578
43. MacNeill C, French R, Evans T, Wessels A, Burch JB (2000) Modular regulation of cGATA-5 gene expression in the developing heart and gut. Dev Biol 217(1):62–76
44. Crispino JD, Lodish M, Thurberg B, Litovsky SH, Collins T, Molkentin J, Orkin SH (2001) Proper coronary vascular development and heart morphogenesis depend on interaction of GATA-4 with FOG cofactors. Genes Dev 15(7):839–844
45. Hatcher CJ, Diman NYSG, Kim M-S, Pennisi D, Song Y, Goldstein MM, Mikawa T, Basson CT (2004) A role for Tbx5 in proepicardial cell migration during cardiogenesis. Physiol Genomics 18(2):129–140
46. Greulich F, Farin HF, Schuster-Gossler K, Kispert A (2012) Tbx18 function in epicardial development. Cardiovasc Res 96:476–483
47. Seo S, Kume T (2006) Forkhead transcription factors, Foxc1 and Foxc2, are required for the morphogenesis of the cardiac outflow tract. Dev Biol 296(2):421–436

48. Kume T, Deng K, Hogan BL (2000) Murine forkhead/winged helix genes Foxc1 (Mf1) and Foxc2 (Mfh1) are required for the early organogenesis of the kidney and urinary tract. Development 127(7):1387–1395
49. Kume T, Jiang H, Topczewska JM, Hogan BL (2001) The murine winged helix transcription factors, Foxc1 and Foxc2, are both required for cardiovascular development and somitogenesis. Genes Dev 15(18):2470–2482
50. Aldinger KA, Lehmann OJ, Hudgins L, Chizhikov VV, Bassuk AG, Ades LC, Krantz ID, Dobyns WB, Millen KJ (2009) FOXC1 is required for normal cerebellar development and is a major contributor to chromosome 6p25.3 Dandy-Walker malformation. Nat Genet 41(9):1037–1042
51. Seo S, Singh HP, Lacal PM, Sasman A, Fatima A, Liu T, Schultz KM, Losordo DW, Lehmann OJ, Kume T (2012) Forkhead box transcription factor FoxC1 preserves corneal transparency by regulating vascular growth. Proc Natl Acad Sci USA 109(6):2015–2020
52. Seo S, Fujita H, Nakano A, Kang M, Duarte A, Kume T (2006) The forkhead transcription factors, Foxc1 and Foxc2, are required for arterial specification and lymphatic sprouting during vascular development. Dev Biol 294(2):458–470
53. Mahtab EAF, Wijffels MCEF, Van Den Akker NMS, Hahurij ND, Lie-Venema H, Wisse LJ, Deruiter MC, Uhrin P, Zaujec J, Binder BR, Schalij MJ, Poelmann RE, Gittenberger-De Groot AC (2008) Cardiac malformations and myocardial abnormalities in podoplanin knockout mouse embryos: correlation with abnormal epicardial development. Dev Dyn 237(3):847–857
54. Breiteneder-Geleff S, Soleiman A, Kowalski H, Horvat R, Amann G, Kriehuber E, Diem K, Weninger W, Tschachler E, Alitalo K, Kerjaschki D (1999) Angiosarcomas express mixed endothelial phenotypes of blood and lymphatic capillaries: podoplanin as a specific marker for lymphatic endothelium. Am J Pathol 154(2):385–394
55. Davies R, Moore AW, Schedl A, Bratt E, Myahawa K, Ladomery M, Miles C, Menke A, van Heyningen V, Hastie ND (1999) Multiple roles for the Wilms' tumor suppressor, WT1. Cancer Res 59(7 suppl):1747s–1750s
56. Rudat C, Kispert A (2012) Wt1 and epicardial fate mapping novelty and significance. Circ Res 111:165–169
57. Christoffels VM, Grieskamp T, Norden J, Mommersteeg MTM, Rudat C, Kispert A (2009) Tbx18 and the fate of epicardial progenitors. Nature 458(7240):E8–E9
58. Merki E, Zamora M, Raya A, Kawakami Y, Wang J, Zhang X, Burch J, Kubalak SW, Kaliman P, Izpisua Belmonte JC, Chien KR, Ruiz-Lozano P (2005) Epicardial retinoid X receptor alpha is required for myocardial growth and coronary artery formation. Proc Natl Acad Sci USA 102(51):18455–18460
59. Ma Q, Zhou B, Pu WT (2008) Reassessment of Isl1 and Nkx2-5 cardiac fate maps using a Gata4-based reporter of Cre activity. Dev Biol 323(1):98–104
60. Pérez-Pomares J-M, de la Pompa JL (2011) Signaling during epicardium and coronary vessel development. Circ Res 109(12):1429–1442
61. Katz TC, Singh MK, Degenhardt K, Rivera-Feliciano J, Johnson RL, Epstein JA, Tabin CJ (2012) Distinct compartments of the proepicardial organ give rise to coronary vascular endothelial cells. Dev Cell 22(3):639–650
62. Wessels A, Pérez-Pomares JM (2004) The epicardium and epicardially derived cells (EPDCs) as cardiac stem cells. Anat Rec 276A(1):43–57
63. Bax NAM, Bleyl SB, Gallini R, Wisse LJ, Hunter J, Van Oorschot AAM, Mahtab EAF, Lie-Venema H, Goumans M-J, Betsholtz C, Gittenberger-De Groot AC (2010) Cardiac malformations in Pdgfrα mutant embryos are associated with increased expression of WT1 and Nkx2.5 in the second heart field. Dev Dyn 239(8):2307–2317
64. Mellgren AM, Smith CL, Olsen GS, Eskiocak B, Zhou B, Kazi MN, Ruiz FR, Pu WT, Tallquist MD (2008) Platelet-derived growth factor receptor signaling is required for efficient epicardial cell migration and development of two distinct coronary vascular smooth muscle cell populations. Circ Res 103(12):1393–1401

65. von Gise A, Zhou B, Honor LB, Ma Q, Petryk A, Pu WT (2011) WT1 regulates epicardial epithelial to mesenchymal transition through β-catenin and retinoic acid signaling pathways. Dev Biol 356(2):421–431
66. Vrancken Peeters M-PFM, Gittenberger-De Groot AC, Mentink MMT, Poelmann RE (1999) Smooth muscle cells and fibroblasts of the coronary arteries derive from epithelial-mesenchymal transformation of the epicardium. Anat Embryol 199(4):367–378
67. Poelmann RE, Gittenberger-de Groot AC, Mentink MM, Bökenkamp R, Hogers B (1993) Development of the cardiac coronary vascular endothelium, studied with antiendothelial antibodies, in chicken-quail chimeras. Circ Res 73(3):559–568
68. Zhou B, Ma Q, Rajagopal S, Wu SM, Domian I, Rivera-Feliciano J, Jiang D, von Gise A, Ikeda S, Chien KR, Pu WT (2008) Epicardial progenitors contribute to the cardiomyocyte lineage in the developing heart. Nature 454(7200):109–113
69. Cai C-L, Martin JC, Sun Y, Cui L, Wang L, Ouyang K, Yang L, Bu L, Liang X, Zhang X, Stallcup WB, Denton CP, McCulloch A, Chen J, Evans SM (2008) A myocardial lineage derives from Tbx18 epicardial cells. Nature 454(7200):104–108
70. Smart N, Bollini S, Dubé KN, Vieira JM, Zhou B, Davidson S, Yellon D, Riegler J, Price AN, Lythgoe MF, Pu WT, Riley PR (2011) De novo cardiomyocytes from within the activated adult heart after injury. Nature 474(7353):640–644
71. Zhou B, Honor LB, Ma Q, Oh J-H, Lin R-Z, Melero-Martin JM, von Gise A, Zhou P, Hu T, He L, Wu KH, Zhang H, Zhang Y, Pu WT (2012) Thymosin beta 4 treatment after myocardial infarction does not reprogram epicardial cells into cardiomyocytes. J Mol Cell Cardiol 52(1):43–47
72. Carmeliet P (2005) Angiogenesis in life, disease and medicine. Nature 438(7070):932–936
73. Adams RH, Alitalo K (2007) Molecular regulation of angiogenesis and lymphangiogenesis. Nat Rev Mol Cell Biol 8(6):464–478
74. Cleaver O, Melton DA (2003) Endothelial signaling during development. Nat Med 9(6):661–668
75. Flamme I, Frölich T, Risau W (1997) Molecular mechanisms of vasculogenesis and embryonic angiogenesis. J Cell Physiol 173(2):206–210
76. Risau W (1997) Mechanisms of angiogenesis. Nature 386(6626):671–674
77. Morabito CJ, Kattan J, Bristow J (2002) Mechanisms of embryonic coronary artery development. Curr Opin Cardiol 17(3):235–241
78. Kattan J, Dettman RW, Bristow J (2004) Formation and remodeling of the coronary vascular bed in the embryonic avian heart. Dev Dyn 230(1):34–43
79. Red-Horse K, Ueno H, Weissman IL, Krasnow MA (2010) Coronary arteries form by developmental reprogramming of venous cells. Nature 464(7288):549–553
80. Lavine KJ, White AC, Park C (2006) Fibroblast growth factor signals regulate a wave of Hedgehog activation that is essential for coronary vascular development. Genes Dev 20(12):1651–1666
81. Luttun A, Carmeliet P (2003) De novo vasculogenesis in the heart. Cardiovasc Res 58(2):378–389
82. Tevosian SG, Deconinck AE, Tanaka M, Schinke M, Litovsky SH, Izumo S, Fujiwara Y, Orkin SH (2000) FOG-2, a cofactor for GATA transcription factors, is essential for heart morphogenesis and development of coronary vessels from epicardium. Cell 101(7):729–739
83. Lavine KJ, Yu K, White AC, Zhang X, Smith C, Partanen J, Ornitz DM (2005) Endocardial and epicardial derived FGF signals regulate myocardial proliferation and differentiation in vivo. Dev Cell 8(1):85–95
84. Grieskamp T, Rudat C, Lüdtke THW, Norden J, Kispert A (2011) Notch signaling regulates smooth muscle differentiation of epicardium-derived cells. Circ Res 108(7):813–823
85. del Monte G, Casanova JC, Guadix JA, MacGrogan D, Burch JBE, Pérez-Pomares J-M, de la Pompa JL (2011) Differential Notch signaling in the epicardium is required for cardiac inflow development and coronary vessel morphogenesis. Circ Res 108(7):824–836
86. Grant RT (1926) Development of the cardiac coronary vessels in the rabbit. Heart 13:261–271

87. Goldsmith JB, Butler HW (1973) The development of the cardiac-coronary circulatory system. Am J Anat 60:185–201
88. Conte G, Pellegrini A (1984) On the development of the coronary arteries in human embryos, stages 14–19. Anat Embryol 169(2):209–218
89. Bogers AJ, Gittenberger-de Groot AC, Dubbeldam JA, Huysmans HA (1988) The inadequacy of existing theories on development of the proximal coronary arteries and their connexions with the arterial trunks. Int J Cardiol 20(1):117–123
90. Bogers AJ, Gittenberger-de Groot AC, Poelmann RE, Péault BM, Huysmans HA (1989) Development of the origin of the coronary arteries, a matter of ingrowth or outgrowth? Anat Embryol 180(5):437–441
91. Waldo KL, Willner W, Kirby ML (1990) Origin of the proximal coronary artery stems and a review of ventricular vascularization in the chick embryo. Am J Anat 188(2):109–120
92. Männer J (1999) Does the subepicardial mesenchyme contribute myocardioblasts to the myocardium of the chick embryo heart? A quail-chick chimera study tracing the fate of the epicardial primordium. Anat Rec 255(2):212–226
93. Riley PR, Smart N (2011) Vascularizing the heart. Cardiovasc Res 91(2):260–268
94. Tammela T, Alitalo K (2010) Lymphangiogenesis: molecular mechanisms and future promise. Cell 140(4):460–476
95. Aselli G (1627) De lactibus sive lacteis venis. J.B. Bidellium, Mediolani
96. Sabin F (1902) On the origin of the lymphatic system from the veins, and the development of the lymph hearts and thoracic duct in the pig. Am J Anat 1:367–389
97. Huntington GS, McClure CFW (1910) The anatomy and development of the jugular lymph sac in the domestic cat (Felis domestica). Am J Anat 10:177–311
98. Srinivasan RS, Dillard ME, Lagutin OV, Lin F-J, Tsai S, Tsai M-J, Samokhvalov IM, Oliver G (2007) Lineage tracing demonstrates the venous origin of the mammalian lymphatic vasculature. Genes Dev 21(19):2422–2432
99. Yaniv K, Isogai S, Castranova D, Dye L, Hitomi J, Weinstein BM (2006) Live imaging of lymphatic development in the zebrafish. Nat Med 12(6):711–716
100. Yang Y, García-Verdugo JM, Soriano-Navarro M, Srinivasan RS, Scallan JP, Singh MK, Epstein JA, Oliver G (2012) Lymphatic endothelial progenitors bud from the cardinal vein and intersomitic vessels in mammalian embryos. Blood 120(11):2340–2348
101. Okuda KS, Astin JW, Misa JP, Flores MV, Crosier KE, Crosier PS (2012) Lyve1 expression reveals novel lymphatic vessels and new mechanisms for lymphatic vessel development in zebrafish. Development 139:2381–2391
102. Schneider M, Othman-Hassan K, Christ B, Wilting J (1999) Lymphangioblasts in the avian wing bud. Dev Dyn 216(4–5):311–319
103. Wilting J, Papoutsi M, Schneider M, Christ B (2000) The lymphatic endothelium of the avian wing is of somitic origin. Dev Dyn 217(3):271–278
104. Wilting J, Aref Y, Huang R, Tomarev SI, Schweigerer L, Christ B, Valasek P, Papoutsi M (2006) Dual origin of avian lymphatics. Dev Biol 292(1):165–173
105. Ny A, Koch M, Schneider M, Neven E, Tong RT, Maity S, Fischer C, Plaisance S, Lambrechts D, Héligon C, Terclavers S, Ciesiolka M, Kälin R, Man WY, Senn I, Wyns S, Lupu F, Brändli A, Vleminckx K, Collen D, Dewerchin M, Conway EM, Moons L, Jain RK, Carmeliet P (2005) A genetic Xenopus laevis tadpole model to study lymphangiogenesis. Nat Med 11:998–1004
106. Buttler K, Kreysing A, von Kaisenberg CS, Schweigerer L, Gale N, Papoutsi M, Wilting J (2006) Mesenchymal cells with leukocyte and lymphendothelial characteristics in murine embryos. Dev Dyn 235(6):1554–1562
107. Lee S, Kang J, Yoo J, Ganesan SK, Cook SC, Aguilar B, Ramu S, Lee J, Hong Y-K (2009) Prox1 physically and functionally interacts with COUP-TFII to specify lymphatic endothelial cell fate. Blood 113(8):1856–1859
108. François M, Caprini A, Hosking B, Orsenigo F, Wilhelm D, Browne C, Paavonen K, Karnezis T, Shayan R, Downes M, Davidson T, Tutt D, Cheah KSE, Stacker SA, Muscat GEO, Achen MG, Dejana E, Koopman P (2008) Sox18 induces development of the lymphatic vasculature in mice. Nature 456(7222):643–647

109. Wigle JT, Harvey N, Detmar M, Lagutina I, Grosveld G, Gunn MD, Jackson DG, Oliver G (2002) An essential role for Prox1 in the induction of the lymphatic endothelial cell phenotype. EMBO J 21(7):1505–1513
110. Srinivasan S, Geng X, Yang Y, Wang Y, Mukatira S, Studer M, Porto MPR, Lagutin OV, Oliver G (2010) The nuclear hormone receptor Coup-TFII is required for the initiation and early maintenance of Prox1 expression in lymphatic endothelial cells. Genes Dev 24(7):696
111. Wigle JT, Oliver G (1999) Prox1 function is required for the development of the murine lymphatic system. Cell 98(6):769–778
112. Kaipainen A, Korhonen J, Mustonen T, Van Hinsbergh VW, Fang GH, Dumont D, Breitman M, Alitalo K (1995) Expression of the fms-like tyrosine kinase 4 gene becomes restricted to lymphatic endothelium during development. Proc Natl Acad Sci USA 92(8):3566–3570
113. Karkkainen MJ, Haiko P, Sainio K, Partanen J, Taipale J, Petrova TV, Jeltsch M, Jackson DG, Talikka M, Rauvala H, Betsholtz C, Alitalo K (2004) Vascular endothelial growth factor C is required for sprouting of the first lymphatic vessels from embryonic veins. Nat Immunol 5(1):74–80
114. Jeltsch M, Tammela T, Alitalo K, Wilting JR (2003) Genesis and pathogenesis of lymphatic vessels. Cell Tissue Res 314(1):69–84
115. Uhrin P, Zaujec J, Breuss JM, Olcaydu D, Chrenek P, Stockinger H, Fuertbauer E, Moser M, Haiko P, Fässler R, Alitalo K, Binder BR, Kerjaschki D (2010) Novel function for blood platelets and podoplanin in developmental separation of blood and lymphatic circulation. Blood 115(19):3997–4005
116. Carramolino L, Fuentes J, García-Andrés C, Azcoitia V, Riethmacher D, Torres M (2010) Platelets play an essential role in separating the blood and lymphatic vasculatures during embryonic angiogenesis. Circ Res 106(7):1197–1201
117. Karpanen T, Wirzenius M, Mäkinen T, Veikkola T, Haisma HJ, Achen MG, Stacker SA, Pytowski B, Ylä-Herttuala S, Alitalo K (2006) Lymphangiogenic growth factor responsiveness is modulated by postnatal lymphatic vessel maturation. Am J Pathol 169(2):708–718
118. Maruyama K, Ii M, Cursiefen C, Jackson DG, Keino H, Tomita M, Van Rooijen N, Takenaka H, D'Amore PA, Stein-Streilein J, Losordo DW, Streilein JW (2005) Inflammation-induced lymphangiogenesis in the cornea arises from CD11b-positive macrophages. J Clin Invest 115(9):2363–2372
119. Baluk P, Tammela T, Ator E, Lyubynska N, Achen MG, Hicklin DJ, Jeltsch M, Petrova TV, Pytowski B, Stacker SA, Ylä-Herttuala S, Jackson DG, Alitalo K, McDonald DM (2005) Pathogenesis of persistent lymphatic vessel hyperplasia in chronic airway inflammation. J Clin Invest 115(2):247–257
120. Zhang Q, Lu Y, Proulx ST, Guo R, Yao Z, Schwarz EM, Boyce BF, Xing L (2007) Increased lymphangiogenesis in joints of mice with inflammatory arthritis. Arthritis Res Ther 9(6):R118
121. Kunstfeld R, Hirakawa S, Hong Y-K, Schacht V, Lange-Asschenfeldt B, Velasco P, Lin C, Fiebiger E, Wei X, Wu Y, Hicklin D, Bohlen P, Detmar M (2004) Induction of cutaneous delayed-type hypersensitivity reactions in VEGF-A transgenic mice results in chronic skin inflammation associated with persistent lymphatic hyperplasia. Blood 104(4):1048–1057
122. Kerjaschki D, Regele HM, Moosberger I, Nagy-Bojarski K, Watschinger B, Soleiman A, Birner P, Krieger S, Hovorka A, Silberhumer G, Laakkonen P, Petrova T, Langer B, Raab I (2004) Lymphatic neoangiogenesis in human kidney transplants is associated with immunologically active lymphocytic infiltrates. J Am Soc Nephrol 15(3):603–612
123. Flister M, Wilber A, Hall K, Iwata C, Miyazono K, Nisato R, Pepper M, Zawieja D, Ran S (2010) Inflammation induces lymphangiogenesis through up-regulation of VEGFR-3 mediated by NF-{kappa}B and Prox1. Blood 115(2):418
124. Mounzer RH, Svendsen OS, Baluk P, Bergman CM, Padera TP, Wiig H, Jain RK, McDonald DM, Ruddle NH (2010) Lymphotoxin-alpha contributes to lymphangiogenesis. Blood 116(12):2173–2182
125. Jiang S, Bailey AS, Goldman DC, Swain JR, Wong MH, Streeter PR, Fleming WH (2008) Hematopoietic stem cells contribute to lymphatic endothelium. PLoS One 3(11):e3812
126. Sebzda E, Hibbard C, Sweeney S, Abtahian F, Bezman N, Clemens G, Maltzman JS, Cheng L, Liu F, Turner M, Tybulewicz V, Koretzky GA, Kahn ML (2006) Syk and Slp-76 mutant

mice reveal a cell-autonomous hematopoietic cell contribution to vascular development. Dev Cell 11(3):349–361
127. Kataru RP, Jung K, Jang C, Yang H, Schwendener RA, Baik JE, Han SH, Alitalo K, Koh GY (2009) Critical role of CD11b+ macrophages and VEGF in inflammatory lymphangiogenesis, antigen clearance, and inflammation resolution. Blood 113(22):5650–5659
128. Salven P, Mustjoki S, Alitalo R, Alitalo K, Rafii S (2003) VEGFR-3 and CD133 identify a population of CD34+ lymphatic/vascular endothelial precursor cells. Blood 101(1):168–172
129. Lee JY, Park C, Cho YP, Lee E, Kim H, Kim P, Yun SH, Yoon Y-S (2010) Podoplanin-expressing cells derived from bone marrow play a crucial role in postnatal lymphatic neovascularization. Circulation 122(14):1413–1425
130. Kerjaschki D, Huttary N, Raab I, Regele H, Bojarski-Nagy K, Bartel G, Kröber SM, Greinix H, Rosenmaier A, Karlhofer F, Wick N, Mazal PR (2006) Lymphatic endothelial progenitor cells contribute to de novo lymphangiogenesis in human renal transplants. Nat Med 12(2):230–234
131. Kajiya K, Sawane M, Huggenberger R, Detmar M (2009) Activation of the VEGFR-3 pathway by VEGF-C attenuates UVB-induced edema formation and skin inflammation by promoting lymphangiogenesis. J Invest Dermatol 129(5):1292–1298
132. Huggenberger R, Ullmann S, Proulx ST, Pytowski B, Alitalo K, Detmar M (2010) Stimulation of lymphangiogenesis via VEGFR-3 inhibits chronic skin inflammation. J Exp Med 207(10):2255–2269
133. Sun QN, Wang YF, Guo ZK (2012) Reconstitution of myocardial lymphatic vessels after acute infarction of rat heart. Lymphology 45(2):80–86
134. Wilting J, Buttler K, Schulte I, Papoutsi M, Schweigerer L, Männer J (2007) The proepicardium delivers hemangioblasts but not lymphangioblasts to the developing heart. Dev Biol 305(2):451–459
135. Doe CQ, Chu-LaGraff Q, Wright DM, Scott MP (1991) The prospero gene specifies cell fates in the drosophila central nervous system. Cell 65(3):451–464
136. Chu-Lagraff Q, Wright DM, McNeil LK, Doe CQ (1991) The prospero gene encodes a divergent homeodomain protein that controls neuronal identity in Drosophila. Development Suppl 2:79–85
137. Oliver G, Sosa-Pineda B, Geisendorf S, Spana EP, Doe CQ, Gruss P (1993) Prox 1, a prospero-related homeobox gene expressed during mouse development. Mech Dev 44(1):3–16
138. Wigle JT, Chowdhury K, Gruss P, Oliver G (1999) Prox1 function is crucial for mouse lens-fibre elongation. Nat Genet 21(3):318–322
139. Hope KJ, Cellot S, Ting SB, MacRae T, Mayotte N, Iscove NN, Sauvageau G (2010) An RNAi screen identifies Msi2 and Prox1 as having opposite roles in the regulation of hematopoietic stem cell activity. Cell Stem Cell 7(1):101–113
140. Wilting J, Papoutsi M, Othman-Hassan K, Rodriguez-Niedenführ M, Pröls F, Tomarev SI, Eichmann A (2001) Development of the avian lymphatic system. Microsc Res Tech 55(2):81–91
141. Del Giacco L, Pistocchi A, Ghilardi A (2010) prox1b Activity is essential in zebrafish lymphangiogenesis. PLoS One 5(10):e13170
142. Hong Y-K, Harvey N, Noh Y-H, Schacht V, Hirakawa S, Detmar M, Oliver G (2002) Prox1 is a master control gene in the program specifying lymphatic endothelial cell fate. Dev Dyn 225(3):351–357
143. Kim H, Nguyen VP, Petrova TV, Cruz M, Alitalo K, Dumont DJ (2010) Embryonic vascular endothelial cells are malleable to reprogramming via Prox1 to a lymphatic gene signature. BMC Dev Biol 10:72
144. Oliver G, Srinivasan RS (2008) Lymphatic vasculature development: current concepts. Ann N Y Acad Sci 1131:75–81
145. Risebro CA, Searles RG, Melville AAD, Ehler E, Jina N, Shah S, Pallas J, Hubank M, Dillard M, Harvey NL, Schwartz RJ, Chien KR, Oliver G, Riley PR (2009) Prox1 maintains muscle structure and growth in the developing heart. Development 136(3):495–505

146. Risebro CA, Petchey LK, Smart N, Gomes J, Clark J, Vieira JM, Yanni J, Dobrzynski H, Davidson S, Zuberi Z, Tinker A, Shui B, Tallini YI, Kotlikoff MI, Miquerol L, Schwartz RJ, Riley PR (2012) Epistatic rescue of Nkx2.5 adult cardiac conduction disease phenotypes by prospero-related homeobox protein 1 and HDAC3. Circ Res 111(2):e19–e31
147. Flaht A, Jankowska-Steifer E, Radomska DM, Madej M, Gula G, Kujawa M, Ratajska A (2012) Cellular phenotypes and spatio-temporal patterns of lymphatic vessel development in embryonic mouse hearts. Dev Dyn 241(9):1473–1486
148. Gordon EJ, Rao S, Pollard JW, Nutt SL, Lang RA, Harvey NL (2010) Macrophages define dermal lymphatic vessel calibre during development by regulating lymphatic endothelial cell proliferation. Development 137(22):3899–3910
149. Smith SJ, Mohun TJ (2011) Early cardiac morphogenesis defects caused by loss of embryonic macrophage function in Xenopus. Mech Dev 128(5–6):303–315
150. Francois M, Harvey NL, Hogan BM (2011) The transcriptional control of lymphatic vascular development. Physiology (Bethesda) 26(3):146–155
151. Shin JW, Min M, Larrieu-Lahargue F, Canron X, Kunstfeld R, Nguyen L, Henderson JE, Bikfalvi A, Detmar M, Hong Y-K (2006) Prox1 promotes lineage-specific expression of fibroblast growth factor (FGF) receptor-3 in lymphatic endothelium: a role for FGF signaling in lymphangiogenesis. Mol Biol Cell 17(2):576–584
152. Cao R, Ji H, Feng N, Zhang Y, Yang X, Andersson P, Sun Y, Tritsaris K, Hansen AJ, Dissing S, Cao Y (2012) Collaborative interplay between FGF-2 and VEGF-C promotes lymphangiogenesis and metastasis. Proc Natl Acad Sci 109(39):15894–15899
153. Kazenwadel J, Secker GA, Betterman KL, Harvey NL (2012) In vitro assays using primary embryonic mouse lymphatic endothelial cells uncover key roles for FGFR1 signalling in lymphangiogenesis. PLoS One 7(7):e40497
154. Kubo H, Cao R, Brakenhielm E, Mäkinen T, Cao Y, Alitalo K (2002) Blockade of vascular endothelial growth factor receptor-3 signaling inhibits fibroblast growth factor-2-induced lymphangiogenesis in mouse cornea. Proc Natl Acad Sci 99(13):8868–8873
155. Larrieu-Lahargue F, Welm AL, Bouchecareilh M, Alitalo K, Li DY, Bikfalvi A, Auguste P (2012) Blocking fibroblast growth factor receptor signaling inhibits tumor growth, lymphangiogenesis, and metastasis. PLoS One 7(6):e39540
156. Choi I, Lee S, Chung HK, Lee YS, Kim KE, Choi D, Park EK, Yang D, Ecoiffier T, Monahan J, Chen W, Aguilar B, Lee HN, Yoo J, Koh CJ, Chen L, Wong AK, Hong Y-K (2012) 9-Cis retinoic acid promotes lymphangiogenesis and enhances lymphatic vessel regeneration: therapeutic implications of 9-Cis retinoic acid for secondary lymphedema. Circulation 125(7):872–882
157. Kang J, Yoo J, Lee S, Tang W, Aguilar B, Ramu S, Choi I, Otu H, Shin J, Dotto G, Koh C, Detmar M, Hong Y-K (2010) An exquisite cross-control mechanism among endothelial cell fate regulators directs the plasticity and heterogeneity of lymphatic endothelial cells. Blood 116(1):140
158. Geudens I, Herpers R, Hermans K, Segura I, Ruiz de Almodovar C, Bussmann J, De Smet F, Vandevelde W, Hogan BM, Siekmann A, Claes F, Moore JC, Pistocchi AS, Loges S, Mazzone M, Mariggi G, Bruyere F, Cotelli F, Kerjaschki D, Noel A, Foidart J-M, Gerhardt H, Ny A, Langenberg T, Lawson ND, Duckers HJ, Schulte-Merker S, Carmeliet P, Dewerchin M (2010) Role of delta-like-4/Notch in the formation and wiring of the lymphatic network in zebrafish. Arterioscler Thromb Vasc Biol 30(9):1695–1702
159. Bochmann L, Sarathchandra P, Mori F, Lara-Pezzi E, Lazzaro D, Rosenthal N (2010) Revealing new mouse epicardial cell markers through transcriptomics. PLoS One 5(6):e11429
160. Gittenberger-De Groot AC, Winter EM, Poelmann RE (2010) Epicardium-derived cells (EPDCs) in development, cardiac disease and repair of ischemia. J Cell Mol Med 14(5):1056–1060
161. Zhou B, Honor LB, He H, Ma Q, Oh J-H, Butterfield C, Lin R-Z, Melero-Martin JM, Dolmatova E, Duffy HS, Gise AV, Zhou P, Hu YW, Wang G, Zhang B, Wang L, Hall JL, Moses MA, McGowan FX, Pu WT (2011) Adult mouse epicardium modulates myocardial injury by secreting paracrine factors. J Clin Invest 121(5):1894–1904

Index

G. Karunamuni (ed.), *The Cardiac Lymphatic System: An Overview*,
DOI 10.1007/978-1-4614-6774-8,

MIX
Papier aus verantwortungsvollen Quellen
Paper from responsible sources
FSC® C105338

If you have any concerns about our products,
you can contact us on
ProductSafety@springernature.com

In case Publisher is established outside the EU,
the EU authorized representative is:
Springer Nature Customer Service Center GmbH
Europaplatz 3, 69115 Heidelberg, Germany

Printed by Libri Plureos GmbH
in Hamburg, Germany